ULTRALEARNING

ULTRALEARNING

Master Hard Skills, Outsmart the Competition, and Accelerate Your Career

SCOTT H. YOUNG
Foreword by James Clear

HARPER
BUSINESS

An Imprint of HarperCollins*Publishers*

FIRST EDITION

Designed by Bonni Leon-Berman
Illustrations on page 20 courtesy of the author
Illustrations on page 188 © Rebecca Lawson

Library of Congress Cataloging-in-Publication Data

Names: Young, Scott, author.
Title: Ultralearning : timeless techniques for mastering hard skills / Scott Young.
Description: New York City : HarperBusiness, [2019] | "This book, originally appearing under the name Nevelj zsenit! was published in Hungarian, without an English translation". | Includes bibliographical references and index. | Summary: "Future-proof your career and maximize your competitive advantage by becoming an Ultralearner --the skill necessary to stay relevant, reinvent yourself, and adapt to whatever the workplace throws your way"-- Provided by publisher.
Identifiers: LCCN 2019021544 | ISBN 9780062852687 (hardcover)
Subjects: LCSH: Career development. | Vocational guidance.
Classification: LCC HF5381 .Y698 2019 | DDC 650.1--dc23
LC record available at https://lccn.loc.gov/2019021544

20 21 22 23   LSC   10 9 8 7 6

To Zorica

CONTENTS

FOREWORD

My relationship with Scott Young began in mid-2013. On July 10, I sent him an email asking if he wanted to set up a call for the following month. We had met at a conference a few days earlier, and I was hoping he would be willing to continue the conversation.

"Possibly," he replied. "I'll be in Spain then, and the language-learning focus of my upcoming project may take precedence."

It wasn't the response I was hoping for, but it seemed reasonable. Managing calls while traveling internationally can be tricky, and I understood if he wanted to wait until he returned. However, I quickly found out that he would not be returning anytime soon, and it was not the time change nor a spotty internet connection that would postpone our conversation.

No, it would be hard to catch up with Scott because he was planning to speak no English *for an entire year*.

Thus began my introduction to Scott Young and his commitment to ultralearning. Over the next twelve months, I would trade sporadic emails with Scott as he traveled to Spain, Brazil, China, and Korea, and proceeded to become conversational in each of the respective languages along the way. He was true to his word: it was not until the following summer in 2014 that we carved out time to catch up regularly and began chatting with each other every few months.

I was always excited for my calls with Scott—primarily for selfish reasons. One of my core interests as a writer is the science of how to build good habits and break bad ones. Someone like

Scott, who had so clearly mastered his own habits, was exactly the type of person who could teach me a thing or two. And that's precisely what happened. I can scarcely remember finishing a call with Scott and not learning something during the previous hour.

That's not to say his insight took me by surprise. Scott had already been on my radar by the time we met at that conference in 2013. He had catapulted to internet fame one year prior by learning the entire MIT undergraduate computer science curriculum and passing all of the final tests in less than a year—four years' worth of classes in under twelve months. I had seen the TEDx Talk summarizing his experience, and I read a few of his articles on learning and self-improvement before tracking him down at the conference.

The idea of taking on an ambitious project—like studying MIT's undergraduate curriculum in one year or learning a new language every three months—is inspirational to many people. I certainly found these bold projects fascinating. But there was something else about Scott's projects that resonated with me on a deeper level: he had a bias toward action.

This is something I have always appreciated about Scott's approach and something I believe you will appreciate as a reader of this book. He isn't focused on simply soaking up knowledge. He is committed to putting that knowledge to use. Approaching learning with an intensity and commitment to action is a hallmark of Scott's process. This approach speaks to me, in part, because I see similar patterns in my own life and career. Some of my most meaningful experiences have been the result of intense self-directed learning.

Although I didn't know the word *ultralearning* at the time, one of my first ultralearning projects was photography. In late 2009,

I moved to Scotland for a few months. It was my first time living abroad, and given the beautiful scenery throughout the Scottish Highlands, I figured I should buy a decent camera. What I hadn't expected, however, was that I would fall in love with the process of taking photos. What followed was one of the most creative periods of my life.

I learned photography through a variety of methods. I studied the portfolios of famous photographers. I scouted locations and searched for compelling perspectives. But, most of all, I learned through one simple method: I took over 100,000 photos that first year. I never enrolled in a photography class. I didn't read books on how to become a better photographer. I just committed to relentless experimentation. This "learning by doing" approach embodies one of my favorite chapters in this book and Scott's third principle of ultralearning: directness.

Directness is the practice of learning by directly doing the thing you want to learn. Basically, it's improvement through active practice rather than through passive learning. The phrases *learning something new* and *practicing something new* may seem similar, but these two methods can produce profoundly different results. Passive learning creates knowledge. Active practice creates skill.

This is a point that Scott more fully clarifies and refines in chapter 6: directness leads to skill development. You can research the best instructions on the bench press technique, but the only way to build strength is to practice lifting weights. You can read all of the bestselling sales books, but the only way to actually get customers is to practice making sales calls. Learning can be very useful, of course, but the danger is that the act of soaking up new facts can be disconnected from the process of refining a new skill. You can know every fact about an industry and still lack real-world expertise because you haven't practiced the craft.

Scott understands the difficulty of actually learning new skills. I respect him not only for the quality of his writing but also for the simple fact that he is a practitioner of his own ideas. I can't say enough about how important this is: he has skin in the game. Many ideas sound brilliant on paper but fail in the real world. As the saying goes, "In theory, there is no difference between theory and practice. But in practice, there is."*

As for my photography quest, it didn't take long for my commitment to direct practice to pay off. A few months after I bought my camera, I traveled to Norway and ventured above the Arctic Circle to capture an image of the aurora borealis. Not long afterward, I was named a finalist for Travel Photographer of the Year thanks to that image of the Northern Lights. It was a surprising outcome, but also a testament to how much progress you can make during a short but intense period of learning.

I never pursued a career as a photographer. It was an ultra-learning project I did for fun and personal satisfaction. But a few years later, right around the time I first met Scott, I began another period of intense learning with a more utilitarian outcome in mind: I wanted to be an entrepreneur, and I figured writing would be one path that could get me there.

Once again, I had selected a domain where I had little formal experience. I had no entrepreneurs in my family, and I had

---

* Credit for this quote has been given to a handful of people over the years, but I believe the earliest source is from 1882 when a student named Benjamin Brewster wrote in the *Yale Literature Magazine*, "I heard no more, for I was lost in self-reproach that I had been the victim of 'vulgar error.' But afterwards, a kind of haunting doubt came over me. What does his lucid explanation amount to but this, that in theory there is no difference between theory and practice, while in practice there is?"

taken only a single college English class. But as I read through *Ultralearning*, I was startled to find that Scott explained, in nearly step-by-step fashion, the process I followed to go from unproven entrepreneur to bestselling author.

Principle #1: Metalearning—I started by examining other popular bloggers and authors. Their methods helped me to create a map for what I needed to do to become a successful writer.

Principle #2: Focus—I went full-time as a writer nearly from the start. Aside from a few freelance projects I took on to pay the bills, the vast majority of my time was spent reading and writing.

Principle #3: Directness—I learned writing by *writing*. I set a schedule for myself to write a new article every Monday and Thursday. Over the first two years, I produced more than 150 essays.

Principle #4: Drill—I systematically broke down each aspect of writing articles—the headline, the introductory sentence, the transitions, the storytelling, and more—and put together spreadsheets filled with examples of each segment. Then I set about testing and refining my ability to perform each small aspect of the larger task.

Principle #6: Feedback—I personally emailed nearly all of my first ten thousand subscribers to say hello and to ask for feedback on my writing. It didn't scale, but it taught me a lot in the beginning.

. . . and so on.

My point is that Scott's method works. By following the techniques he lays out in this book, I was able to build a writing career, create a successful business, and, ultimately, write a *New York Times* bestselling book. When I released *Atomic Habits*, it was the culmination of years of work centered around the process of ultralearning.

I think it's easy to hear stories about writing a bestselling book or learning four languages in a year and think, "That's for other people." I disagree. Learning something valuable and doing it fast doesn't have to be confined to some narrow set of geniuses. It's a process anyone can embrace. It's just that most people never do it because they never had a playbook to show them how. Until now.

There are good reasons to pursue ultralearning—whether you are conducting a project for personal or professional interests.

First, deep learning provides a sense of purpose in life. Developing skills is meaningful. It feels good to get good at something. Ultralearning is a path to prove to yourself that you have the ability to improve and to make the most of your life. It gives you the confidence that you can accomplish ambitious things.

Second, deep learning is how you get outsized returns. The simple truth is most people will never intensely study your area of interest. Doing so—even if it's just for a few months—will help you stand out. And once you stand out, you can get a better job, negotiate for a higher salary or more free time, network with more interesting people, and otherwise level up your personal and professional life. Ultralearning helps you develop leverage that you can use elsewhere.

Finally, deep learning is possible. Paul Graham, the famous entrepreneur and investor, once noted, "In many fields a year of focused work plus caring a lot would be enough."* Similarly, I think most people would be surprised by what they could accomplish with a year (or a few months) of focused learning. The process of intense self-directed learning can fashion skills you never thought you could develop. Ultralearning can help you

---

* Paul Graham, "How to Be an Expert in a Changing World," December 2014, http://www.paulgraham.com/ecw.html?viewfullsite=1.

fulfill your potential, and that is perhaps the best reason of all to pursue it.

The truth is, despite the success of my writing and photography pursuits, these projects were haphazard. I did them intensely but without guidance or direction. I made a lot of mistakes. I wish I had this book when I was starting out. I can only imagine how much wasted time and energy I would have saved.

Ultralearning is a fascinating and inspiring read. Scott has compiled a gold mine of actionable strategies for learning anything faster. His effort is now your gain. I hope you enjoy this book as much as I did, and, most important, I hope you use these ideas to accomplish something ambitious and exciting in your own life. With the stories and strategies Scott shares in this book, you will have the knowledge. All that is left is to take action.

—*James Clear*

ULTRALEARNING

CHAPTER I

Can You Get an MIT Education Without Going to MIT?

Only a few hours left. I caught myself glancing out the window as the early-morning light glittered off the buildings in front of me. It was a crisp fall day, surprisingly sunny for a famously rainy city. Well-dressed men carried briefcases and fashionable women pulled miniature dogs beneath my eleventh-story vantage point. Buses dragged reluctant commuters into town one last time before the weekend. The city might have been rousing from its slumber, but I had been awake since before dawn.

Now is not the time for daydreaming, I reminded myself and shifted my attention back to the half-finished math problems scribbled on the notebook in front of me. "Show that $\iint_R \text{curl} F \cdot \hat{n} \, dS = 0$ for any finite part of the unit sphere . . ." the problem began. The class was Multivariate Calculus for the Massachusetts Institute of Technology. The final exam would start soon, and I had little time left to prepare. *What was curl, again . . . ?* I closed my eyes and tried to form a picture of the problem in my head. *There's a sphere. I know that.* I conjured a bright red ball in my mind's eye, floating in empty space. *Now what was ñ? The ñ stands for normal*, I reminded myself, meaning an arrow that points straight up from the surface. My red ball became furry, with hairlike vectors standing straight at all ends. *But what about curl?* My imagination turned to waves of tiny arrows pulsating in a vast sea. Curl marked the eddies, swirling around in little loops. I thought again to my furry, red ball with the static-charged hairdo. My fuzzy sphere had no whorls, so there must not be any curl, I reasoned. *But how do I prove it?* I scratched down some equations. *Better double-check it.* My mental pictures were clear, but my symbol manipulation was a lot sloppier. There wasn't much time left, and every second of preparation counted. I needed to grind through as many problems as possible before time ran out.

That was nothing unusual for an MIT student. Tricky equations, abstract concepts, and difficult proofs are all a normal part of one of the most prestigious educations in math and science in the world. Except that I was not an MIT student. In fact, I had never even been to Massachusetts. All of this was taking place in my bedroom, twenty-five hundred miles away in Vancouver, Canada. And although an MIT student typically covers the entirety of multivariate calculus over a semester, I had started only five days before.

THE MIT CHALLENGE

I have never attended MIT. Instead, my college days were spent studying business at the University of Manitoba, a middle-ranked Canadian school I could actually afford. After graduating with a bachelor of commerce, I felt as though I had picked the wrong major. I wanted to be an entrepreneur and so had studied business, thinking that would be the best route to becoming my own boss. Four years later, I discovered that a business major was largely a finishing school for entrants into the world of big corporations, gray suits, and standard operating procedures. Computer science, in contrast, was a major where you actually learned to make things. Programs, websites, algorithms, and artificial intelligence were what had interested me in entrepreneurship in the first place, and I was struggling to decide what to do about it.

I could go back to school, I thought. Enroll again. Spend another four years working toward a second degree. But taking out student loans and giving up a half decade of my life to repeat the bureaucracy and rules of college didn't seem very appealing. There had to be a better way to learn what I wanted.

Around that time, I stumbled across a class taught at MIT and posted online. It had fully recorded lectures, assignments, and quizzes; even the actual exams used in the real class with the solution keys were provided. I decided to try taking the class. To my surprise, I found that the class was much better than most of the classes I had paid thousands of dollars to attend in university. The lectures were polished, the professor was engaging, and the material was fascinating. Digging further, I could see that this wasn't the only class MIT offered for free. MIT had uploaded the materials from hundreds of different classes. I wondered if this could be the solution to my problem. If anyone could learn the content of an MIT class for free, would it be possible to learn the content of an entire degree?

Thus began almost six months of intense research into a project I named the MIT Challenge. I looked up the actual MIT curriculum for computer science undergrads. I matched and compared the list with the resources MIT offered online. Unfortunately, that was a lot easier said than done. MIT's OpenCourseWare, the platform used for uploading class material, had never been intended as a substitute for attending the school. Some classes simply weren't offered and needed to be swapped out. Others had such scant material that I wondered if they would even be possible to complete. Computation Structures, one of the required courses, which taught how to build a computer from scratch using circuits and transistors, had no recorded lectures or assigned textbook. To learn the class content, I would have to decipher abstract symbols written on a slideshow meant to accompany the lecture. Missing materials and ambiguous evaluation criteria meant that doing every class exactly as an MIT student would was out of the question. However, a simpler approach might work: just try to pass the final exams.

This focus on final exams later expanded to include programming projects for the classes that had them. These two criteria formed the skeleton of an MIT degree, covering most of the knowledge and skills I wanted to learn, with none of the frills. No mandatory attendance policy. No due dates on assignments. The final exams could be taken whenever I was ready and retaken with an alternate exam if I happened to fail one. Suddenly what had initially seemed like a disadvantage—not having physical access to MIT—became an advantage. I could approximate the education of an MIT student for a fraction of the cost, time, and constraints.

Exploring this possibility further, I even did a test class using the new approach. Instead of showing up to prescheduled lectures, I watched downloaded videos for the class at twice the normal speed. Instead of meticulously doing each assignment and waiting weeks to learn my results, I could test myself on the material one question at a time, quickly learning from my mistakes. Using these and other methods, I found I could scrape through a class in as little as a week's time. Doing some quick calculations and adding some room for error, I decided it might be possible to tackle the remaining thirty-two classes in under a year.

Although it began as a personal quest, I started to see that there were bigger implications beyond my little project. Technology has made learning easier than ever, yet tuition costs are exploding. A four-year degree used to be an assurance of a decent job. Now it is barely a foot in the door. The best careers demand sophisticated skills that you're unlikely to stumble upon by chance. Not just programmers but managers, entrepreneurs, designers, doctors, and nearly every other profession is rapidly accelerating the knowledge and skills required, and many are struggling to keep up. In the back of my mind, I was interested not only in computer science but in seeing if there might be a new way to master the skills needed in work and life.

As my attention drifted once more to the scene developing outside my window, I thought about how all this had started. I thought about how I wouldn't be attempting my odd little experiment at all had it not been for a chance encounter with an intense, teetotaling Irishman on another continent almost three years earlier.

FLUENT IN THREE MONTHS?

"My problem isn't with the French—just Parisians," Benny Lewis vented to me in an Italian restaurant in the heart of Paris. Lewis was vegetarian, not always easy to accommodate in a country famous for steak tartare and foie gras. Eating a plate of penne arrabbiata, a favorite he had picked up while working in a youth hostel in Italy, Lewis spoke in fluent French, not minding much if any of the locals overheard his complaints. His discontent stemmed from a particularly dreary year working as a *stagiaire* in an engineering firm in Paris. He had found it hard to adjust to the notorious job demands and social life in France's biggest city. Still, he thought, perhaps he shouldn't be too critical. It was that experience, after all, that had led him to leave his life as an engineer and travel around the world learning languages.

I had been introduced to Lewis during a moment of personal frustration. I was living in France as part of a student exchange program. I had left home with high hopes of ending the year speaking effortless French, but things didn't seem to be turning out that way. Most of my friends spoke to me in English, including the French ones, and it was starting to feel as though one year wouldn't be enough.

I complained about this state of affairs to a friend from home; he told me about a guy he had heard of who traveled from country to country, challenging himself to learn a language in three months. "Bullshit," I said, with more than a hint of envy. Here I had been struggling to chat with people after months of immersion, and this guy was challenging himself to do so after only three months. Despite my skepticism, I knew I needed to meet Lewis to see if he understood something about learning languages that I didn't. An email and a train ride later, and Lewis and I were meeting face-to-face.

"Always have a challenge," Lewis told me as he continued with his life advice, now guiding me on a postlunch tour of central Paris: Lewis's earlier feelings about Paris were starting to soften, and as we walked from the Notre Dame to the Louvre, his mood turned nostalgic about his days in the city. His strong opinions and passions, I would later learn, not only fueled his desire to take on ambitious challenges but could also get him into trouble. He was once detained by Brazilian federal police after an immigration officer overheard him cursing her in Portuguese to friends outside when she had denied him a visa extension. The irony was that his visa had been denied because she didn't believe his Portuguese could be so good from such a short stay, and she suspected him of secretly trying to immigrate to Brazil outside the terms of his tourist visa.

As we continued to walk, now on the grounds in front of the Eiffel Tower, Lewis explained his approach: Start speaking the very first day. Don't be afraid to talk to strangers. Use a phrasebook to get started; save formal study for later. Use visual mnemonics to memorize vocabulary. What struck me were not the methods but the boldness with which he applied them. While I had timidly been trying to pick up some French, worrying about saying the wrong things and being embarrassed by my insufficient vocabulary, Lewis was fearless, diving straight into conversations and setting seemingly impossible challenges for himself.

That approach had served him well. He was already fluent in Spanish, Italian, Gaelic, French, Portuguese, Esperanto, and English and had recently reached a conversational level while staying in the Czech Republic for three months. But it was his newest challenge he was planning that intrigued me the most: fluency in German after just three months.

It wasn't, strictly speaking, Lewis's first time with German. He had taken German classes for five years in high school and had briefly visited Germany twice before. However, like many of the students who spent time learning a language in school, he still couldn't speak it. He admitted with embarrassment, "I couldn't even order breakfast in German if I wanted to." Still, the unused knowledge built up from classes taken over a decade earlier would probably make his challenge easier than starting from scratch. To compensate for the reduced difficulty, Lewis decided to raise the stakes.

Normally, he challenged himself to reach the equivalent of a B2 level in a language after three months. The B2 level—the fourth out of six levels beginning A1, A2, B1, and so on—is described by the Common European Framework of Reference for Languages (CEFR) as upper intermediate, allowing the speaker to "interact with a degree of fluency and spontaneity that makes regular interaction with native speakers quite possible without strain for either party." However, for his German challenge, Lewis decided to go for the highest exam level offered: C2. This level represents a complete mastery of the language. To reach a C2 level, the learner must "understand with ease virtually everything heard or read" and "express themselves spontaneously, very fluently and precisely, differentiating finer shades of meaning even in the most complex situations." The Goethe-Institut, which administers the exam, recommends at least 750 hours of instruction, not including extensive practice outside the classroom, to reach this benchmark.

A few months later, I heard back from Lewis about his project. He had missed his goal of passing the C2 exam by a hair. He had passed four of five criteria for his exam but had failed the listening comprehension section. "I spent too much time listening to the radio," he chastised himself. "I should have done more active listening practice." Fluency in three months of intensive practice had eluded him, although he had come surprisingly close. In the seven years after my first encounter with the Irish polyglot, he has gone on to attempt his three-month challenge

in half a dozen more countries, adding to his linguistic reper-
toire some Arabic, Hungarian, Mandarin Chinese, Thai, American Sign Language, and even Klingon (the invented *Star Trek*
language).

What I didn't realize at the time but understand now was that
Lewis's accomplishments weren't all that rare. In the space of
linguistic feats alone, I have encountered hyperpolyglots who
speak forty-plus languages, adventurer-anthropologists who can
start speaking previously unknown languages after a few hours
of exposure, and many other travelers, like Lewis, who hop from
tourist visa to tourist visa, mastering new languages. I also saw
that this phenomenon of aggressive self-education with incredible results wasn't restricted to languages alone.

HOW ROGER CRAIG GAMED *JEOPARDY!*

"What is *The Bridge on the River Kwai*?" Roger Craig hastily scribbled the question on his screen. Despite first fumbling over the
legibility of the film title's final word, Craig was correct. He had
won $77,000—the highest single-day winnings in *Jeopardy!* his-
tory at the time. Craig's victory wasn't a fluke. He broke records
again, amassing nearly two hundred thousand dollars, the highest ever five-game winning streak. Such a feat would be remarkable on its own, but what was more incredible was how he did it.
Reflecting on the moment, Craig says, "My first thought wasn't
'Wow, I just won seventy-seven thousand dollars.' It was 'Whoa,
my site really worked.'"

How do you study for a test that can ask any question? That
was the essential problem Craig faced as he prepared to compete. *Jeopardy!* is famous for stumping home audiences with trivia

questions that can ask about anything from Danish kings to Damocles. Thus the great champions of *Jeopardy!* tend to be brainy know-it-alls who have spent a lifetime amassing the huge library of factual knowledge needed to spit out answers on any topic. Studying for *Jeopardy!* might feel like an impossible task, as you would need to study almost every conceivable subject. Craig's solution, however, was to rethink the process of acquiring knowledge itself. To do that, he built a website.

"Everybody that wants to succeed at a game is going to practice the game," Craig contends. "You can practice haphazardly, or you can practice efficiently." To amass the wide-ranging trivia needed to break records, he decided to be ruthlessly analytical about how he acquired knowledge. A computer scientist by trade, he decided to start off by downloading the tens of thousands of questions and answers from every *Jeopardy!* game ever aired. He tested himself on those during his free time for months, and then, as it became clear that he was going to go on television, he switched to aggressively quizzing himself on the questions fulltime. He then applied text-mining software to categorize the questions into different topics, such as art history, fashion, and science. He used data visualization to map out his strengths and weaknesses. The text-mining software separated the different topics, which he visualized as different circles. The position of any given circle on his graph showed how good he was in that topic—higher meant he knew more about that topic. The size of the circle indicated how frequent that topic was. Bigger circles were more common and thus better choices for further study. Beneath the variety and randomness in the show, he started to uncover hidden patterns. Some clues in the show are "Daily Doubles," which allow a contestant to double his or her score, or lose it all. These extremely valuable clues may seem randomly placed, but having the entire *Jeopardy!* archives at his fingertips, Craig found that their position followed trends. One could hunt out the valuable Doubles by hopping between categories and focusing on high-point clues, breaking the conventional approach to the show of sticking within a single category until it was completed.

Craig also found trends within the types of questions asked. Although *Jeopardy!* could conceivably ask questions on any topic, the format of the game is designed to entertain a home audience, not to challenge competitors. Following this reasoning, Craig found that he could get away with studying the best-known trivia within a category, rather than digging deep into any particular direction. If a subject was specialized, he knew the answers would be geared toward the best-known examples. By analyzing his own weakness on archival questions, he could see which topics he needed to study more to be competitive. For example, he found that he was weak on fashion and focused on studying that topic more deeply.

Using analytics to figure out what to study was only the first step. From there, Craig employed spaced-repetition software to maximize his efficiency. Spaced-repetition software is an advanced flash card algorithm first developed by the Polish researcher Piotr Woźniak in the 1980s. Woźniak's algorithm was designed to optimally time when you need to review material in order to remember it. Given a large database of facts, most people will forget what they learn first, needing to remind themselves of it again and again for it to stick. The algorithm fixes this problem by calculating the optimal time for reviewing each fact so you don't waste energy overdrilling the same information, but also so you don't forget what you've already learned. This tool allowed Craig to efficiently memorize the thousands of facts he would need for his later victory.

Although the show airs only one episode per day, *Jeopardy!* tapes five episodes at a time. Craig was coming back to his hotel room after winning five games straight, and he couldn't sleep. He said, "You can simulate the game, but you can't simulate winning two hundred thousand dollars in five hours and setting the single-day record on a game show you've wanted to be on since you were twelve years old." Combining unorthodox tactics and aggressive analysis, he had gamed the game show and won.

Roger Craig wasn't the only person I found who had seen his fortunes change as a result of aggressive self-education. I didn't know it at the time, but in 2011, the same year my MIT Challenge would begin, Eric Barone was starting his own obsession. Unlike mine, however, his efforts would extend for nearly five years and require mastering many completely different skills.

FROM MINIMUM WAGE TO MILLIONAIRE

Eric Barone had just graduated from the University of Washington Tacoma with a degree in computer science when he thought, *Now's my chance.* He had decided that he wanted to make his own video games and that now, before he got comfortable in a salaried programming job, was his opportunity to do something about it. He already had his inspiration. He wanted his game to pay homage to *Harvest Moon*, a charming Japanese series of games in which the player must build a successful farm: grow crops, raise animals, explore the countryside, and form relationships with other villagers. "I loved that game," he said about his childhood experience with the title. "But it could have been so much better." He knew that if he didn't follow through with his own vision, that improved version would never be a reality.

Developing a commercially successful video game isn't easy. AAA game companies budget hundreds of millions of dollars and employ thousands of people on their top titles. The talent required is similarly broad. Game development requires programming, visual art, musical composition, story writing, game design, and dozens more skills, depending on the genre and style of game developed. The breadth of skills required makes game development much harder for smaller teams than other art forms such as music, writing, or visual arts. Even highly talented independent game developers generally have to collaborate with a few people to span all the skills required. Eric Barone, however, decided to work on his game entirely alone.

Deciding to work alone came from a personal commitment to his vision and an indefatigable self-confidence that he could finish the game. "I like to have complete control over my own vision," he explained, saying that it might have been "impossible to find people who were on the same page" regarding the design. However, that choice meant that he would need to become proficient in game programming, music composition, pixel art, sound design, and story writing. More than just a game design project, Barone's odyssey would entail mastering each aspect of game design itself.

Pixel art was Barone's biggest weakness. This style of art harkens back to the earlier era of video games when rendering graphics was difficult to do on slow computers. Pixel art is not done with fluid lines or photorealistic textures. Instead, a compelling image must be created by placing pixels, the colored dots that make up computer graphics, one at a time—painstaking and difficult work. A pixel artist must convey movement, emotion, and life from a grid of colored squares. Barone liked to doodle and draw, but that didn't prepare him for the difficulty. He had to learn this skill "completely from scratch." Getting his art skills to a commercial level wasn't easy. "I must have done most of the artwork three to five times over," he said. "For the character portraits, I did those at least ten times."

Barone's strategy was simple but effective. He practiced by working directly on the graphics he wanted to use in his game. He critiqued his own work and compared it to art he admired. "I tried to break it down scientifically," he explained. "I would ask myself, 'Why do I like this? Why don't I like that?'" when looking at other artists' work. He supplemented his own practice by reading about pixel art theory and finding tutorials that could fill gaps in his knowledge. When he encountered a difficulty in his art, he broke it down: "I asked, 'What goal do I want to reach?' and then 'How might I get there?'" At some point in his work on the game, he felt his colors were too dull and boring. "I wanted the colors to pop," he said. So he researched color theory and intensively studied other artists to see how they used colors to make things visually interesting.

Pixel art was just a single aspect Barone had to learn. He also composed all of the music for his game, redoing it from scratch more than once to make sure it met his high expectations. Whole sections of the game mechanics were developed and scrapped when they failed to meet his rigorous standards. This process of practicing directly and redoing things allowed him to get steadily better at all of the aspects of game design. Although it lengthened the time it took to complete the game, it also enabled his finished product to compete with games created by an army of specialized artists, programmers, and composers.

Throughout the five-year development process, Barone avoided seeking employment as a computer programmer. "I didn't want to get involved in something substantial," he said. "I wouldn't have had the time, and I wanted to give game development my best shot." Instead, he worked as a theater usher, earning minimum wage so that he wouldn't get distracted. His meager earnings from his job, combined with support from his girlfriend, allowed Barone to get by as he focused on his passion.

That passion and dedication to mastery paid off. Barone released *Stardew Valley* in February 2016. The game quickly became a surprise hit, outselling many of the big-studio titles offered on the computer game platform Steam. Across multiple platforms, Barone estimates that within the first year of its release, *Stardew Valley* had sold well over 3 million copies. In months, he went from an unknown designer earning minimum wage to a millionaire named one of *Forbes*' "30 Under 30" stars within game development. His dedication to mastering the skills involved played no small part in that success. Destructoid, in its review of *Stardew Valley*, described the artwork as "incredibly endearing and beautiful." Barone's commitment to his vision and aggressive self-education had paid off handsomely.

THE MIT CHALLENGE AND BEYOND

Back in my cramped apartment, I was grading my calculus exam. It was tough, but it looked as though I had passed. I was relieved, but it wasn't a time to relax. Next Monday, I would be starting all over again, with a new course, and I still had almost a year to go.

As the calendar changed, so did my strategies. I switched from trying to do a single class in several days to spending a month doing three to four classes in parallel. I hoped that would spread the learning over a longer period of time and reduce some of the negative effects of cramming. As I made more progress, I also slowed down. My first few classes were done with aggressive haste so I could stay on schedule to meet my self-imposed deadline. After it seemed likely that I could finish, I was able to shift from studying sixty hours per week to studying thirty-five to forty. Finally, in September 2012, less than twelve months after I had begun, I finished the final class.

Completing the project was eye-opening for me. For years, I had thought the only way to learn things deeply was to push through school. Finishing this project taught me not only that this assumption was false but that this alternate path could be more fun and exciting. In university, I had often felt stifled, trying to stay awake during boring lectures, grinding through busywork assignments, forcing myself to learn things I had no interest in just to get the grade. Because this project was my own vision and design, it rarely felt painful, even if it was often challenging. The subjects felt alive and exciting, rather than stale chores to be completed. For the first time ever, I felt I could learn anything I wanted to with the right plan and effort. The possibilities were endless, and my mind was already turning toward learning something new.

Then I got a message from a friend: "You're on the front page of Reddit, you know." The internet had found my project, and it was generating quite a discussion. Some liked the idea but doubted its usefulness: "It's sad that employers won't really treat this in the same way as a degree, even if he has the same amount (or more) knowledge than a graduate does." One user claiming to be the head of R&D for a software company disagreed: "This is the type of person I want. I really do not care if you have a degree or not." The debate raged. Had I actually done it or not? Would I be able to get a job as a programmer after this? Why try to do this in a year? Was I crazy?

The initial surge of attention led to other requests. An employee at Microsoft wanted to set me up for a job interview. A new startup asked me to join its team. A publishing house in China offered me a book deal to share some studying tips with beleaguered Chinese students. However, those weren't the reasons I had done the project. I was already happy working as a writer online, which had supported me financially throughout my project and would continue to do so afterward. My goal for the project wasn't to get a job but to see what was possible. After just a few months of finishing my first big project, ideas for new ones were already bubbling up inside my head.

I thought of Benny Lewis, my first example in this strange world of intense self-education. Following his advice, I had eventually reached an intermediate level of French. It had been hard work, and I was proud that I had been able to push against my initial difficulty of being surrounded by a bubble of English speakers to learn enough French to get by. However, after finishing my MIT experience, I was injected with a new confidence I hadn't had in France. What if I didn't make the mistake I made last time? What if, instead of forming a group of English-speaking friends and struggling to pop out of that bubble once my French was good enough, I emulated Benny Lewis and dived straight into immersion from the very first day? How much better could I be, if as in my MIT Challenge, I held nothing back and optimized everything around learning a new language as intensely and effectively as possible?

As luck would have it, around that time my roommate was planning on going back to grad school and wanted some time off to travel first. We'd both been saving, and if we pooled our resources and were frugal in how we planned our trip, we figured we might be able to do something exciting. I told him about my experiences in France, both of learning French and of secretly believing that much more was possible. I told him about the social bubble that had formed when I had arrived without speaking the language and how difficult it had been to break out of it later. What if, instead of just hoping you'd practice enough, you don't give yourself an escape route? What if you commit to speaking only the language you're trying to learn from the first moment you step off the plane? My friend was skeptical. He had seen me study MIT classes for a year from across our apartment. My sanity was still an open question, but he wasn't as confident in his own ability. He wasn't sure he could do it, although he was willing to give it a shot, as long as I didn't have any expectations of him to succeed.

That project, which my friend and I titled "The Year Without English," was simple. We'd go to four countries, three months each. The plan in each country was straightforward: no speaking English, either with each other or with anyone we'd meet, from the first day. From there we'd see how much we could learn before our tourist visas ran out and we were pushed to a new destination.

Our first stop was Valencia, Spain. We had just landed in the airport when we encountered our first obstacle. Two attractive British girls came up to us, asking for directions. We looked at each other and awkwardly sputtered out the little Spanish we knew, pretending we didn't speak any English. They didn't understand us and asked us again, now in an exasperated tone. We stumbled over some more Spanish and, believing we couldn't speak English, they walked away in frustration. Already, it seemed, not speaking English was having unintended consequences. Despite that inauspicious beginning, our Spanish ability grew even faster than I had anticipated. After two months in Spain, we were interacting in Spanish beyond what I had achieved in an entire year of partial immersion in France. We would go to our tutor in the morning, study a little at home, and spend the rest of the day hanging out with friends, chatting at restaurants, and soaking up the Spanish sun. My friend, despite his earlier doubts, was also a convert to this new approach to learning things. Although he didn't care to study grammar and vocabulary as aggressively as I did, by the end of our stay, he too was integrating seamlessly into life in Spain. The method worked far better than we had hoped, and we were now believers.

We continued the trip, going to Brazil to learn Portuguese, China to learn Mandarin, and South Korea to learn Korean. Asia proved a far harder task than Spain or Brazil. In our preparation, we had assumed those languages would be only a little more difficult than the European ones, although it turned out that they were much harder. As a result, our no-English rule was starting to crack, although we still applied it as much as we could. Even if our Mandarin and Korean didn't reach the same level of ability after a short stay, it was still enough to make friends, travel, and converse with people on a variety of topics. At the end of our year, we could confidently say we spoke four new languages.

Having seen the same approach work for academic computer science and language-learning adventures, I was slowly becoming convinced that it could be applied to much more. I had enjoyed drawing as a kid, but like most people's attempts, any faces I drew looked awkward and artificial. I had always admired people who could quickly sketch a likeness, whether it be street-side caricaturists to professional portrait painters. I wondered if the same approach to learning MIT classes and languages could also apply to art.

I decided to spend a month improving my ability to draw faces. My main difficulty, I realized, was in placing the facial features properly. A common mistake when drawing faces, for instance, is putting the eyes too far up the head. Most people think they sit in the top two-thirds of the head. In truth, they're more typically halfway between the top of the head and the chin. To overcome these and other biases, I did sketches based on pictures. Then I would take a photo of the sketch with my phone and overlay the original image on top of my drawing. Making the photo semi-transparent allowed me to see immediately whether the head was too narrow or wide, the lips too low or too high or whether I had put the eyes in the right spot. I did this hundreds of times, employing the same rapid feedback strategies that had served me well with MIT classes. Applying this and other strategies, I was able to get a lot better at drawing portraits in a short period of time (see below).

DAY 1   DAY 30

UNCOVERING THE ULTRALEARNERS

On the surface, projects such as Benny Lewis's linguistic adventures, Roger Craig's trivia mastery, and Eric Barone's game development odyssey are quite different. However, they represent instances of a more general phenomenon I call *ultralearning*.* As I dug deeper, I found more stories. Although they differed in the specifics of what had been learned and why, they shared a common thread of pursuing extreme, self-directed learning projects and employed similar tactics to complete them successfully.

Steve Pavlina is an ultralearner. By optimizing his university schedule, he took a triple course load and completed a computer science degree in three semesters. Pavlina's challenge long predated my own experiment with MIT courses and was one of the first inspirations that showed me compressing learning time might be possible. Done without the benefit of free online classes, however, Pavlina attended California State University, Northridge, and graduated with actual degrees in computer science and mathematics.

Diana Jaunzeikare embarked on an ultralearning project to replicate a PhD in computational linguistics. Benchmarking Carnegie Mellon University's doctoral program, she wanted to not only take classes but also conduct original research. Her project had started because going back to academia to get a real doctorate would have meant leaving the job she loved at Google. Like many other ultralearners before her, Jaunzeikare's project

---

* Technically, the term *ultralearning* was first used by Cal Newport, in his headline to an article I wrote for his website about my recently completed MIT Challenge, which he titled "Mastering Linear Algebra in 10 Days: Astounding Experiments in Ultra-Learning."

was an attempt to fill a gap in education when formal alternatives didn't fit with her lifestyle.

Facilitated by online communities, many ultralearners operate anonymously, their efforts observable only by unverifiable forum postings. One such poster at Chinese-forums.com, who goes only by the username Tamu, extensively documented his process of studying Chinese from scratch. Devoting "70–80+ hours each week" over four months, he challenged himself to pass the HSK 5, China's second highest Mandarin proficiency exam.

Other ultralearners shed the conventional structures of exams and degrees altogether. Trent Fowler, starting in early 2016, embarked on a yearlong effort to become proficient in engineering and mathematics. He titled it the STEMpunk Project, a play on the STEM fields of science, technology, engineering, and mathematics he wanted to cover and the retrofuturistic steampunk aesthetic. Fowler split his project into modules. Each module covered a particular topic, including computation, robotics, artificial intelligence, and engineering, but was driven by hands-on projects instead of copying formal classes.

Every ultralearner I encountered was unique. Some, like Tamu, preferred punishing, full-time schedules to meet harsh, self-imposed deadlines. Others, like Jaunzeikare, managed their projects on the side while maintaining full-time jobs and work obligations. Some aimed at the recognizable benchmarks of standardized exams, formal curricula, and winning competitions. Others designed projects that defied comparison. Some specialized, focusing exclusively on languages or programming. Others desired to be true polymaths, picking up a highly varied set of skills.

Despite their idiosyncrasies, the ultralearners had a lot of shared traits. They usually worked alone, often toiling for months and years without much more than a blog entry to announce their efforts. Their interests tended toward obsession. They were aggressive about optimizing their strategies, fiercely debating the merits of esoteric concepts such as interleaving practice, leech thresholds, or keyword mnemonics. Above all, they cared about learning. Their motivation to learn pushed them to tackle intense projects, even if it often came at the sacrifice of credentials or conformity.

The ultralearners I met were often unaware of one another. In writing this book, I wanted to bring together the common principles I observed in their unique projects and in my own. I wanted to strip away all the superficial differences and strange idiosyncrasies and see what learning advice remains. I also wanted to generalize from their extreme examples something an ordinary student or professional can find useful. Even if you're not ready to tackle something as extreme as the projects I've described, there are still places where you can adjust your approach based on the experience of ultralearners and backed by the research from cognitive science.

Although the ultralearners are an extreme group of people, this approach to things holds potential for normal professionals and students. What if you could create a project to quickly learn the skills to transition to a new role, project, or even profession? What if you could master an important skill for your work, as Eric Barone did? What if you could be knowledgeable about a wide variety of topics, like Roger Craig? What if you could learn a new language, simulate a university degree program, or become good at something that seems impossible to you right now?

Ultralearning isn't easy. It's hard and frustrating and requires stretching outside the limits of where you feel comfortable. However, the things you can accomplish make it worth the effort.

Let's spend a moment trying to see what exactly ultralearning is and how it differs from the most common approaches to learning and education. Then we can examine what the principles are that underlie all learning, to see how ultralearners exploit them to learn faster.

CHAPTER II

Why Ultralearning Matters

What exactly is ultralearning? While my introduction to the eclectic group of intense autodidacts started with seeing examples of unusual learning feats, to go forward we need something more concise. Here's an imperfect definition:

> ULTRALEARNING: A strategy for acquiring skills and knowledge that is both self-directed and intense.

First, ultralearning is a strategy. A strategy is not the only solution to a given problem, but it may be a good one. Strategies also tend to be well suited for certain situations and not others, so using them is a choice, not a commandment.

Second, ultralearning is self-directed. It's about how you make decisions about what to learn and why. It's possible to be a completely self-directed learner and still decide that attending a particular school is the best way to learn something. Similarly, you could "teach yourself" something on your own by mindlessly following the steps outlined in a textbook. Self-direction is about who is in the driver's seat for the project, not about where it takes place.

Finally, ultralearning is intense. All of the ultralearners I met took unusual steps to maximize their effectiveness in learning. Fearlessly attempting to speak a new language you've just started to practice, systematically drilling tens of thousands of trivia questions, and iterating through art again and again until it is perfect is hard mental work. It can feel as though your mind is at its limit. The opposite of this is learning optimized for fun or convenience: choosing a language-learning app because it's entertaining, passively watching trivia show reruns on television so you don't feel stupid, or dabbling instead of serious practice. An intense method might also produce a pleasurable state of flow, in which the experience of challenge absorbs your focus and you lose track of time. However, with ultralearning, deeply and effectively learning things is always the main priority.

This definition covers the examples I've discussed so far, but in some ways it is unsatisfyingly broad. The ultralearners I've met have a lot more overlapping qualities than this minimal definition implies. This is why in the second part of the book I'll discuss deeper principles that are common in ultralearning and how they can enable some impressive achievements. Before that, however, I want to explain why I think ultralearning matters—because although the examples of ultralearning may seem eccentric, the benefits of this approach to learning are deep and practical.

THE CASE FOR ULTRALEARNING

It's obvious that ultralearning isn't easy. You'll have to set aside time from your busy schedule in order to pursue something that will strain you mentally, emotionally, and possibly even physically. You'll be forced to face down frustrations directly without retreating into more comfortable options. Given this difficulty, I think it's important to articulate clearly why ultralearning is something you should seriously consider.

The first reason is for your work. You already expend much of your energy working to earn a living. In comparison, ultralearning is a small investment, even if you went so far as to temporarily make it a full-time commitment. However, rapidly learning hard skills can have a greater impact than years of mediocre striving on the job. Whether you want to change careers, take on new challenges, or accelerate your progress, ultralearning is a powerful tool.

The second reason is for your personal life. How many of us have dreams of playing an instrument, speaking a foreign language, becoming a chef, writer, or photographer? Your deepest moments of happiness don't come from doing easy things; they come from realizing your potential and overcoming your own limiting beliefs about yourself. Ultralearning offers a path to master those things that will bring you deep satisfaction and self-confidence.

Although the motivation behind ultralearning is timeless, let's start by looking at why investing in mastering the art of learning hard things quickly is going to become even more important to your future.

ECONOMICS: AVERAGE IS OVER

In the words of the economist Tyler Cowen, "Average is over." In his book of the same title, Cowen argues that because of increased computerization, automation, outsourcing, and regionalization, we are increasingly living in a world in which the top performers do a lot better than the rest.

Driving this effect is what is known as "skill polarization."
It's well known that income inequality has been increasing in
the United States over the last several decades. However, this
description ignores a more subtle picture. The MIT economist
David Autor has shown that instead of inequality rising across
the board, there are actually two different effects: inequality rising at the top and lowering at the bottom. This matches Cowen's
thesis of average being over, with the middle part of the income
spectrum being compressed into the bottom and stretched out at
the top. Autor identifies the role that technology has had in creating this effect. The advance of computerization and automation
technologies has meant that many medium-skilled jobs—clerks,
travel agents, bookkeepers, and factory workers—have been
replaced with new technologies. New jobs have arisen in their
place, but those jobs are often one of two types: either they are
high-skilled jobs, such as engineers, programmers, managers,
and designers, or they are lower-skilled jobs such as retail workers, cleaners, or customer service agents.

Exacerbating the trends caused by computers and robots are
globalization and regionalization. As medium-skilled technical work is outsourced to workers in developing nations, many
of those jobs are disappearing at home. Lower-skilled jobs,
which often require face-to-face contact or social knowledge in
the form of cultural or language abilities, are likely to remain.
Higher-skilled work is also more resistant to shipping overseas
because of the benefits of coordination with management and
the market. Think of Apple's tagline on all of its iPhones: "Designed in California. Made in China." Design and management
stay; manufacturing goes. Regionalization is a further extension
of this effect, with certain high-performing companies and cities
making outsized impacts on the economy. Superstar cities such

as Hong Kong, New York, and San Francisco have dominating effects on the economy as firms and talent cluster together to take advantage of proximity.

This paints a picture that might either be bleak or hopeful, depending on your response to it. Bleak, because it means that many of the assumptions embedded in our culture about what is necessary to live a successful, middle-class lifestyle are quickly eroding. With the disappearance of medium-skilled jobs, it's not enough to get a basic education and work hard every day in order to succeed. Instead, you need to move into the higher-skilled category, where learning is constant, or you'll be pushed into the lower-skilled category at the bottom. Underneath this unsettling picture, however, there is also hope. Because if you can master the personal tools to learn new skills quickly and effectively, you can compete more successfully in this new environment. That the economic landscape is changing may not be a choice any of us has control over, but we can engineer our response to it by aggressively learning the hard skills we need to thrive.

EDUCATION: TUITION IS TOO HIGH

The accelerating demand for high-skilled work has increased the demand for college education. Except instead of expanding into education for all, college has become a crushing burden, with skyrocketing tuition costs making decades of debt a new normal for graduates. Tuition has increased far faster than the rate of inflation, which means that unless you are well poised to translate that education into a major salary increase, it may not be worth the expense.

Many of the best schools and institutions fail to teach many of the core vocational skills needed to succeed in the new high-skilled jobs. Although higher education has traditionally been a place where minds were shaped and characters developed, those lofty goals seem increasingly out of touch with the basic financial realities facing new graduates. Therefore, even for those who do go to college, there are very often skill gaps between what was learned in school and what is needed to succeed. Ultralearning can fill some of those gaps when going back to school isn't an affordable option.

Rapidly changing fields also mean that professionals need to constantly learn new skills and abilities to stay relevant. While going back to school is an option for some, it's out of reach for many. Who has the ability to put their life on hold for years as they wade through classes that may or may not end up covering the situations they actually need to deal with? Ultralearning, because it is directed by learners themselves, can fit into a wider variety of schedules and situations, targeting exactly what you need to learn without the waste.

Ultimately, it doesn't matter if ultralearning is a suitable replacement for higher education. In many professions, having a degree isn't just nice, it's legally required. Doctors, lawyers, and engineers all require formal credentials to even start doing the job. However, those same professionals don't stop learning when they leave school, and so the ability to teach oneself new subjects and skills remains essential.

TECHNOLOGY: NEW FRONTIERS IN LEARNING

Technology exaggerates both the vices and the virtues of humanity. Our vices are made worse because now they are downloadable, portable, and socially transmissible. The ability to distract or delude yourself has never been greater, and as a result we are facing crises of both privacy and politics. Though those dangers are real, there is also opportunity created in their wake. For those who know how to use technology wisely, it is the easiest time in history to teach yourself something new. An amount of information vaster than was held by the Library of Alexandria is freely accessible to anyone with a device and an internet connection. Top universities such as Harvard, MIT, and Yale are publishing their best courses for free online. Forums and discussion platforms mean that you can learn in groups without ever leaving your home.

Added to these new advantages is software that accelerates the act of learning itself. Consider learning a new language, such as Chinese. A half century ago, learners needed to consult cumbersome paper dictionaries, which made learning to read a nightmare. Today's learner has spaced-repetition systems to memorize vocabulary, document readers that translate with the tap of a button, voluminous podcast libraries offering endless opportunities for practice, and translation apps that smooth the transition to immersion. This rapid change in technology means that many of the best ways of learning old subjects have yet to be invented or rigorously applied. The space of learning possibilities is immense, just waiting for ambitious autodidacts to come up with new ways to exploit it.

Ultralearning does not require new technology, though. As I will discuss in the chapters to come, the practice has a long history, and many of the most famous minds could be described as having applied some version of it. However, technology offers an incredible opportunity for innovation. There are still many ways to learn things that we have yet to fully explore. Perhaps certain learning tasks could be made far easier or even obsolete, with the right technical innovation. Aggressive and efficiency-minded ultralearners will be the first to master them.

ACCELERATE, TRANSITION, AND RESCUE YOUR CAREER WITH ULTRALEARNING

The trends toward skill polarization in the economy, skyrocketing tuition, and new technology are all global. But what does ultralearning actually look like for an individual? I believe there are three main cases in which this strategy for quickly acquiring hard skills can apply: accelerating the career you have, transitioning to a new career, and cultivating a hidden advantage in a competitive world.

To see how ultralearning can accelerate the career you already have, consider Colby Durant. After graduating from college, she started work at a web development firm but wanted to make faster progress. She took on an ultralearning project to learn copywriting. After taking the initiative and showing her boss what she could do, she was able to get a promotion. By choosing a valuable skill and focusing on quickly developing proficiency, you can accelerate your normal career progression.

Learning is often the major obstacle to transitioning to the career you want to have. Vishal Maini, for instance, was comfortable in his marketing role in the tech world. But he dreamed of being more closely involved with artificial intelligence research. Unfortunately, that was a deep technical skill set that he hadn't acquired. Through a careful six-month ultralearning project, however, he was able to develop strong enough skills that he could switch fields and get a job working in the field he wanted.

Finally, an ultralearning project can augment the other skills and assets you've cultivated in your work. Diana Fehsenfeld worked as a librarian for years in her native New Zealand. Facing government cutbacks and rapid technologization of her field, she was worried that her professional experience might not be enough to keep up. As a result, she undertook two ultralearning projects, one to learn statistics and the programming language R and another on data visualization. Those skills were in demand in her industry, and adding them to her background as a librarian gave her the tools to go from bleak prospects to being indispensable.

BEYOND BUSINESS: THE CALL TO ULTRALEARNING

Ultralearning is a potent skill for dealing with a changing world. The ability to learn hard things quickly is going to become increasingly valuable, and thus it is worth developing to whatever extent you can, even if it requires some investment first.

Professional success, however, was rarely the thing that motivated the ultralearners I met—including those who ended up making the most money from their new skills. Instead it was a compelling vision of what they wanted to do, a deep curiosity, or even the challenge itself that drove them forward. Eric Barone didn't pursue his passion in solitude for five years to become a millionaire but because he wanted the satisfaction of creating something that perfectly matched his vision. Roger Craig didn't want to go on *Jeopardy!* to win prize money but to push himself to compete on the show he had loved since he was a child. Benny Lewis didn't learn languages to become a technical translator, or later a popular blogger, but because he loved traveling and interacting with the people he met along the way. The best ultralearners are those who blend the practical reasons for learning a skill with an inspiration that comes from something that excites them.

There's an added benefit to ultralearning that transcends even the skills one learns with it. Doing hard things, particularly things that involve learning something new, stretches your self-conception. It gives you confidence that you might be able to do things that you couldn't do before. My feeling after my MIT Challenge wasn't just a deepened interest in math and computer science but an expansion in possibility: If I could do this, what else could I do that I was hesitant to try before? Learning, at its core, is a broadening of horizons, of seeing things that were previously invisible and of recognizing capabilities within yourself that you didn't know existed. I see no higher justification for pursuing the intense and devoted efforts of the ultralearners I've described than this expansion of what is possible. What could you learn if you took the right approach to make it successful? Who could you become?

WHAT ABOUT TALENT? THE TERENCE TAO PROBLEM

Terence Tao is smart. By age two, he had taught himself to read. At age seven, he was taking high school math classes. By seventeen, he had finished his master's thesis. It was titled "Convolution Operators Generated by Right-Monogenic and Harmonic Kernels." After that, he got a PhD from Princeton, won the coveted Fields Medal (called by some the "Nobel Prize for mathematics"), and is considered to be one of the best mathematical minds alive today. Though many mathematicians are extreme specialists—rare orchids adapted to thrive only on a particular branch of the mathematical tree—Tao is phenomenally diverse. He regularly collaborates with mathematicians and makes important contributions to distant fields. This virtuosity caused one colleague to liken his ability to "a leading English-language novelist suddenly producing the definitive Russian novel."

What's more, there doesn't seem to be an obvious explanation for his feats. He was precocious, certainly, but his success in mathematics didn't come from aggressively overbearing parents pushing him to study. His childhood was filled playing with his two younger brothers, inventing games with the family's Scrabble board and mah-jongg tiles, and drawing imaginary maps of fantasy terrain. Normal kid stuff. Nor does he seem to have a particularly innovative studying method. As noted in his profile in the *New York Times*, he coasted on his intelligence so far that, upon reaching his PhD, he fell back "on his usual test-prep strategy: last-minute cramming." Although that approach faltered once he reached the pinnacle of his field, the fact that he breezed through classes for so long points to a powerful mind rather than some unique strategy. *Genius* is a word thrown around too casually, but in Tao's case the label certainly sticks.

Terence Tao and other naturally gifted learners present a major challenge for the universality of ultralearning. If people like Tao can accomplish so much without aggressive or inventive studying methods, why should we bother investigating the habits and methods of other impressive learners? Even if the feats of Lewis, Barone, or Craig don't reach the level of Tao's brilliance, perhaps their accomplishments also are due to some hidden intellectual ability that normal people lack. If this were so, ultralearning might be something interesting to examine but not something you could actually replicate.

PUTTING TALENT ASIDE

What role does natural talent play? How can we examine what causes someone's success when the shadow of intelligence and innate gifts looms over us? What do stories like Tao's mean for mere mortals who just want to improve their capacity to learn?

The psychologist K. Anders Ericsson argues that particular types of practice can change most attributes necessary for becoming an expert-level performer with the exception of the innate traits of height and body size. Other researchers are less optimistic about the malleability of our natures. Many argue that a substantial proportion, perhaps most, of our intelligence is genetically derived. If intelligence comes mostly from genes, why not use that to explain ultralearning instead of ultralearners' use of a more effective method or strategy? Tao's success in mathematics doesn't seem to be owed to something easily replicable by normal human beings, so why assume that any of the ultralearners are any different?

I take a middle position between those two extremes. I think that natural talents exist and that they undoubtedly influence the results we see (especially at extreme levels, as in the case of Tao). I also believe that strategy and method matter, too. Throughout this book, I will cover science showing how making changes to how you learn can impact your effectiveness. Each of the principles is something that, if applied appropriately, will make you a better learner regardless of whether your starting point is dull or brilliant.

My approach in telling stories for this book, therefore, will not be to try to determine what the sole cause is of someone's intellectual success. Not only is this impossible, but it isn't particularly useful. Instead, I'm going to use stories and anecdotes to illustrate and isolate what are the most practical and useful things you can do to improve how you learn. The ultralearners I mention should serve as exemplars you can use to see how a principle applies in practice, not a guarantee that you can achieve an identical result with identical effort.

FINDING TIME FOR ULTRALEARNING

Another doubt that may have formed in your mind when reading so far is asking how you'll find time to do these intensive learning projects. You may worry that this advice won't apply to you because you already have work, school, or family commitments that prevent you from throwing yourself into learning full-time.

In practice, however, this usually isn't a problem. There are three main ways you can apply the ideas of ultralearning, even if you have to manage other commitments and challenges in your life: new part-time projects, learning sabbaticals, and reimagining existing learning efforts.

The first way is by pursuing ultralearning part-time. The most dramatic examples of learning success tend to be those where the ultralearner put impressive amounts of time into the project. Spending fifty hours a week on a project will accomplish more than spending five hours a week on it, even if the efficiency is the same, and thus the most captivating stories usually involve heroic schedules. Though this makes for good storytelling, it's actually unnecessary when it comes to pursuing your own ultralearning projects. The core of the ultralearning strategy is intensity and a willingness to prioritize effectiveness. Whether this happens on a full-time schedule or just a couple hours per week is completely up to you. As I'll discuss in chapter 10, a spread-out schedule may even be more efficient in terms of long-term memory. Whenever you read about an intensive schedule in this book, feel free to adapt it to your own situation, taking a more leisurely pace while employing the same ruthlessly efficient tactics.

The second way is by pursuing ultralearning during gaps in work and school. Many of the people I interviewed did their projects during temporary unemployment, career transitions, semesters off, or sabbaticals. Although these aren't as reliable to plan for, a burst of learning may be perfect for you if you know you're about to have this kind of time off. That was one of my motivations for pursuing my MIT Challenge when I did: I had just graduated, so extending my existing student life another year was easier than pushing it out for four. If I had to do the same project today, I might have done it over a longer period of time, over some evenings and weekends, since my work is less flexible today than it was in that moment of transition from school to working life.

The third way is to integrate ultralearning principles into the time and energy you already devote to learning. Think about the last business book you read or the time you tried to pick up Spanish, pottery, or programming. What about that new software you needed to learn for work? Those professional development hours you need to log to maintain your certification? Ultralearning doesn't have to be an additional activity; it can inform the time you already spend learning. How can you align the learning and studies you already need to do with the principles for maximizing effectiveness?

As in the section on talent, don't let the extreme examples dissuade you from applying the same principles. Everything I will share with you can be customized or integrated into what already exists. What matters is the intensity, initiative, and commitment to effective learning, not the particulars of your timetable.

THE VALUE OF ULTRALEARNING

The ability to acquire hard skills effectively and efficiently is immensely valuable. Not only that, but the current trends in economics, education, and technology are going to exacerbate the difference between those with this skill and those without it. In this discussion, however, I've ignored perhaps the most important question: Ultralearning may be valuable, but is it learnable? Is ultralearning just a description of people with unusual personalities, or does it represent something that someone who wasn't an ultralearner before could actually become?

CHAPTER III

How to Become an Ultralearner

"I'd love to be a guinea pig." It was an email from Tristan de Montebello. I had first met the charming, half-French, half-American musician and entrepreneur seven years earlier, at almost exactly the same time as my fateful encounter with Benny Lewis. With tousled blond hair and a close-cropped beard, he looked like he belonged on a surfboard on some stretch of California coastline. De Montebello was the kind of guy you liked immediately: confident, yet down to earth, with only the vaguest hints of a French accent in his otherwise perfect English. Over the years, we had kept in touch: me with my strange learning experiments; him hopping around the world, going from working with a Parisian startup that made bespoke cashmere sweaters to guitarist, vagabond, and eventually web consultant in Los Angeles, much closer to the beaches that suited him so well. Now he had heard I was writing a book about learning, and he was interested.

The context of his email was that although I had met and documented dozens of people accomplishing strange and intriguing learning feats, the meetings had been largely after the fact. They were people I had met or heard about after their successes, not before; observations of successes, not experiments that generated them. As a result, it was hard to tell exactly how accessible this ultralearning thing was. If you filter through enough pebbles, you're sure to find a few flecks of gold. Was I doing the same thing, scouring for unusual learning projects? Sift through enough people, and you're bound to find some that seem incredible. But if ultralearning had the potential that I imagined it did, it would be nice to find someone before he or she tried a project and watch the results. To test this, I put together a small group of about a dozen people (mostly readers of my blog) who were interested in giving this ultralearning thing a shot. Among them was de Montebello.

BECOMING AN ULTRALEARNER

"Maybe piano?" de Montebello suggested. Although he was interested in the concept of ultralearning, he had no idea which skill he'd like to learn. He had played guitar and been the lead singer for a band. With his musical background, learning to play piano seemed like a relatively safe choice. He had even made a course teaching guitar lessons online, so learning another musical instrument could potentially expand his business. Selfishly, I encouraged him to try learning something farther outside his comfort zone. A musician picking up another instrument didn't seem like the ideal case to study for seeing whether ultralearning could be applied broadly. We threw more ideas around. A week or two later, he decided on public speaking. His background as a musician had given him experience being onstage, but otherwise he had little experience giving speeches. Public speaking is a useful skill, too, he argued, so it would be worth getting better at even if nothing noteworthy came from the effort.

De Montebello had a private motivation for wanting to become good at public speaking. He had given only a handful of speeches in his life, and most of those had been in college. He related to me one example, from when he had gone to give a talk to a dozen people at a web design firm in Paris: "I cringe every time I think back to that." He explained, "I could just tell I wasn't connecting. There were many pieces where I was boring them. There were jokes where I would laugh, because I thought it was funny, but nobody else would." Being a musician, he was surprised "how little of it translated" to public speaking. Still, it was something in which he saw potential value, if he could get good at it. "Public speaking is a metaskill," he feels. It's the kind of skill that assists with other skills: "confidence, storytelling, writing, creativity, interviewing skills, selling skills. It touches on so many different things." With that in mind, he set to work.

FIRST STEPS OF A FLEDGLING ULTRALEARNER

De Montebello had picked his topic, but he wasn't sure exactly how he should learn it. He decided to attend a meeting of Toastmasters International, the organization for learning public speaking. At that point, his story had two doses of luck. The first was that in attendance at the very first meeting for his public speaking project was Michael Gendler. Gendler was a longtime Toastmaster, and de Montebello's combination of charm and obsessive intensity to become good at public speaking convinced him to help coach de Montebello through his project. The second dose of luck was something that de Montebello didn't fully appreciate at the time: he had showed up just ten days before the deadline to be eligible to compete in the World Championship of Public Speaking.

The World Championship of Public Speaking is a competition put on annually by Toastmasters in which members compete, elimination style, starting in individual clubs and going on to larger and larger units of the organization, until a select few make it to the final stage. De Montebello had little more than a week to prepare. Still, the competition provided a potential structure for his ultralearning project, so he went for it, cranking out the six mandatory qualifying speeches in the coming week, finishing the last in the nick of time.

De Montebello practiced obsessively, sometimes speaking twice in one day. He recorded a video of every speech and analyzed it obsessively for flaws. He asked for feedback every time he gave a speech, and he got plenty of it. His coach, Gendler, pushed him far outside of his comfort zone. Once, when faced with the choice between polishing an existing speech and creating a brand-new one from scratch, de Montebello asked what he should do. Gendler's response was to do whichever was scariest for him.

His relentless drive pushed de Montebello into unusual places. He took improv classes to work on his spontaneous delivery. There he learned to trust whatever was in his head and deliver it without hesitation. That kept him from stammering over his words or fearing freezing up onstage. He talked to a friend who works as a Hollywood director to give feedback on his delivery. The director taught de Montebello to give his speech dozens of times in different styles—angry, monotone, screaming, even as a rap—then go back and see what was different from his normal voice. According to de Montebello, that helped break him of the "uncanny valley" that happened when his normal speaking delivery felt slightly unnatural.

Another friend with a background in theater gave him tips on stage presence. He took de Montebello through his speech and showed how each word and sentence indicated movement that could be translated to where he moved on the stage. Instead of standing constricted under the spotlight, de Montebello could now move gracefully and use his body to communicate his message on top of his words. He even gave his speech at a middle school, knowing that seventh graders would give the most ruthless feedback of all. After bombing terribly outside the comfort of Toastmasters, he learned to talk to his audience before going onstage: learn their language and emotions and connect with them. That way, applying all he had learned so far, he could change his speech on the fly, so it would be sure to connect with a new audience. Above all, Gendler pushed him relentlessly. "Make me care," Gendler told him after listening to one of de Montebello's speeches. "I understand why this is important to you, but the audience doesn't care about you. You have to make me care." Diverse advice and voluminous practice would soak those lessons in deeply, allowing de Montebello to quickly surpass his early awkwardness on the stage.

After a month, de Montebello won his area competition, beating out a competitor with two decades of experience in Toastmasters. He won his district and division competitions, too. Finally, less than seven months after he first tried his hand at public speaking, he was going to compete in the World Championships. "There are about thirty thousand people who compete every year," he noted, adding "I'm pretty confident I'm the fastest competitor in history to make it this far, since if I had started ten days later, I couldn't have competed." He made it into the top ten.

FROM SEMIFINALIST TO CAREER CHANGE

"I knew this project was going to be big for me when I started it," de Montebello told me months after his top ten placement in the international competition. "But it was literally life changing. I didn't expect it to actually change my life." Reaching the final competition in the World Championship had been quite a journey, but it was only afterward that he began to realize how much he had learned. "I was learning for this very narrow world of public speaking. It was only after that I realized the depth of all these skills I had worked so much on: storytelling, confidence, communicating."

Friends who heard of de Montebello's success started asking if he could help them work on their own speeches. He and Gendler saw an opportunity to help others improve their public speaking skills. The demand was intense. Authors who command five-figure speaking fees started to approach the duo to see if they could be taught to improve their public speaking the ultralearning way. Soon they had landed their first client, to the tune of twenty thousand dollars. Gendler and de Montebello weren't mercenary; they wanted to focus only on speakers whose message they really believed in. But the fact that they had attracted such high-status clients certainly helped persuade them to make the switch into coaching public speaking full-time. Gendler and de Montebello even decided to name their consultancy UltraSpeaking, as a nod to the strategy that made it all possible.

De Montebello's story ended up being much more dramatic than either of us had initially expected. His initial hope had been that he could practice intensely for a few months, deliver a great speech somewhere, and have it recorded—a nice memento and a new skill—but not that he would reach the status of an international competitor and eventually experience a complete career change. Of the other dozen or so people I spent some time coaching into ultralearning, none offered so dramatic an example. Some dropped out. Life got in the way (or perhaps revealed that they weren't actually as committed as they had initially appeared to be). Others had respectable successes, making significant improvements in learning medicine, statistics, comic book drawing, military history, and yoga, even if they didn't reach de Montebello's degree of success.

What differentiated de Montebello wasn't that he thought he could go from near-zero experience to the finalist for the World Championship in six months. Rather, it was his obsessive work ethic. His goal wasn't to reach some predetermined extreme but to see how far he could go. Sometimes you'll get lucky and embark on a path that will take you quite far. But even the failure mode of ultralearning is usually that you will learn a skill fairly well. Even those who didn't have such dramatic results among the small group I spent time coaching, those who stuck with their project still ended up learning a new skill they cared about. You may not compete in world championships or completely switch careers, but as long as you stick with the process, you're bound to learn something new. What de Montebello's example encapsulated for me was not only that you can become an ultralearner but that such successes are far from being the inevitable consequences of having a particular kind of genius or talent. Had de Montebello focused on piano instead, his experience with giving speeches would probably have remained that one awkward example from his days in Paris.

PRINCIPLES OF BECOMING AN ULTRALEARNER

De Montebello's story illustrates that it's possible to decide to become an ultralearner. But ultralearning isn't a cookie-cutter method. Every project is unique, and so are the methods needed to master it. The uniqueness of ultralearning projects is one of the elements that ties them all together. If ultralearning could be bottled or standardized, it would simply be an intense form of structured education. What makes ultralearning interesting is also what makes it hard to boil down into step-by-step formulas.

This is a difficult challenge, but I'm going to try to sidestep it by focusing on principles first. Principles allow you to solve problems, even those you may have never encountered before, in a way that a recipe or mechanical procedure cannot. If you really understand the principles of physics, for instance, you can solve a new problem simply by working backward. Principles make sense of the world, and even if they don't always articulate exactly how you should solve a particular challenge, they can provide immense guidance. Ultralearning, in my view, works best when you see it through a simple set of principles, rather than trying to copy and paste exact steps or protocols.

The principles of ultralearning are going to be the focus of the second part of this book. In each chapter, I'll introduce a new principle, plus some evidence to back it up both from ultralearning examples and from scientific research. Finally, I'll share possible ways that the principle can manifest itself as specific tactics. These tactics are only a small sample. But they should provide a starting point for you to think creatively about your own ultralearning challenges.

There are nine universal principles that underlie the ultralearning projects described so far. Each embodies a particular aspect of successful learning, and I describe how ultralearners maximize the effectiveness of the principle through the choices they make in their projects. They are:

1. **Metalearning: First Draw a Map.** Start by learning how to learn the subject or skill you want to tackle. Discover how to do good research and how to draw on your past competencies to learn new skills more easily.

2. **Focus: Sharpen Your Knife.** Cultivate the ability to concentrate. Carve out chunks of time when you can focus on learning, and make it easy to just do it.

3. **Directness: Go Straight Ahead.** Learn by doing the thing you want to become good at. Don't trade it off for other tasks, just because those are more convenient or comfortable.

4. **Drill: Attack Your Weakest Point.** Be ruthless in improving your weakest points. Break down complex skills into small parts; then master those parts and build them back together again.

5. **Retrieval: Test to Learn.** Testing isn't simply a way of assessing knowledge but a way of creating it. Test yourself before you feel confident, and push yourself to actively recall information rather than passively review it.

6. **Feedback: Don't Dodge the Punches.** Feedback is harsh and uncomfortable. Know how to use it without letting your ego get in the way. Extract the signal from the noise, so you know what to pay attention to and what to ignore.

7. **Retention: Don't Fill a Leaky Bucket.** Understand what you forget and why. Learn to remember things not just for now but forever.

8. **Intuition: Dig Deep Before Building Up.** Develop your intuition through play and exploration of concepts and skills. Understand how understanding works, and don't recourse to cheap tricks of memorization to avoid deeply knowing things.

9. **Experimentation: Explore Outside Your Comfort Zone.** All of these principles are only starting points. True mastery comes not just from following the path trodden by others but from exploring possibilities they haven't yet imagined.

I organized these nine principles based on my observations of ultralearning projects as well as my own personal experience, referencing, where I could, the vast cognitive science literature. I started with the ultralearners themselves. If one person did something in a certain way, that might be an interesting example, but it might also be an idiosyncrasy of that person. If several people or, better yet, every ultralearner I encountered, did a certain thing in a certain way, it was much stronger evidence that I had stumbled upon a general principle. I then checked those principles against the scientific literature. Are there mechanisms and findings from cognitive science to support the tactics I saw? Better yet, have there been controlled experiments comparing one approach to learning with another? The scientific research supports many of the learning strategies employed by the ultralearners I witnessed. This suggests that ultralearners, with their ruthless focus on efficiency and effectiveness, may have landed on some universal principles in the art of learning.

Beyond principles and tactics is a broader ultralearning ethos. It's one of taking responsibility for your own learning: deciding what you want to learn, how you want to learn it, and crafting your own plan to learn what you need to. You're the one in charge, and you're the one who's ultimately responsible for the results you generate. If you approach ultralearning in that spirit, you should take these principles as flexible guidelines, not as rigid rules. Learning well isn't just about following a set of prescriptions. You need to try things out for yourself, think hard about the nature of the learning challenges you face, and test solutions to overcome them. With that in mind, let's turn to the first ultralearning principle: metalearning.

CHAPTER IV

PRINCIPLE 1

Metalearning

First Draw a Map

> If I have seen further it is by standing on the
> shoulders of giants.
> —*Isaac Newton*

Dan Everett stands in front of a packed auditorium. A stocky man in his early sixties, he speaks slowly and confidently, his smiling face framed by thinning blond hair and a beard. Next to him is a table filled with assorted objects: sticks, stones, leaves, containers, fruit, a pitcher of water. He signals that the demonstration is about to begin.

Entering from a door on the right, a heavyset middle-aged woman with dark brown hair and olive skin approaches the stage. Everett goes up to her and says something in a language she doesn't understand. She looks around, clearly confused, and then replies hesitantly, "Kuti paoka djalou." He tries to repeat what she has just said. There's some stumbling at first, but after one or two more tries, she seems satisfied with his repetition of her reply. He goes to the blackboard and writes, "Kuti paoka djalou ⇨ 'Greeting (?).'" Next he picks up a small stick and points to it. She guesses correctly that he wants to know the name and replies, "ŋkindo." Once again, Everett goes to the blackboard and writes, "ŋkindo ⇨ stick." Next he tries two sticks and gets the same response, "ŋkindo." He then drops the stick, to which the woman says, "ŋkindo paula." The demonstration proceeds, with Everett picking up objects, performing actions, listening to the woman's responses, and recording the results on the blackboard. Soon he's moved past simple naming tasks and starts asking for more complicated sentences: "She drinks the water," "You eat the banana," and "Put the rock in the container." With each new elicitation, he experiments, building new sentences and testing her reaction to see if he is correct. Within half an hour, there are more than two blackboards full of nouns, verbs, pronouns, and phonetic annotations.

Learning dozens of words and phrases in a new language is a good start for the first thirty minutes spent with any language. What makes this feat particularly impressive is that Everett isn't allowed to speak any language he might have in common with the speaker. He can only try to encourage her to say words and phrases and repeat them to try to figure out the language's grammar, pronunciation, and vocabulary. He doesn't even know what language is being spoken.[*]

How can Everett start speaking a new language from scratch, without teachers or translations or even knowing what language he's learning, in half an hour, when most of us struggle to do the

---

[*] The language of the speaker, it turned out, was a dialect of Hmong, spoken in parts of China, Vietnam, and Laos.

same after years of high school Spanish classes? What enables Everett to pick up vocabulary, decode grammar, and pronunciation so much faster than you or I, even with all those additional constraints? Is he a linguistic genius, or is there something else going on?

The answer is our first principle of ultralearning: metalearning.

WHAT IS METALEARNING?

The prefix *meta* comes from the Greek term μετά, meaning "beyond." It typically signifies when something is "about" itself or deals with a higher layer of abstraction. In this case metalearning means learning about learning. Here's an example: If you're learning Chinese characters, you will learn that 火 means "fire." That's regular learning. You may also learn that Chinese characters are often organized by something called radicals, which indicate what kind of thing the character describes. The character 灶, for example which means "stove," has a 火 on the left-hand side to indicate that it has some relationship to fire. Learning this property of Chinese characters is metalearning—not learning about the object of your inquiry itself, in this case words and phrases, but learning about how knowledge is structured and acquired within this subject; in other words, learning how to learn it.

In Everett's case, we can see glimpses of the enormous wealth of metalearning that lies just beneath the surface. "Well, what are some of the things we noticed about this?" Everett asks the audience after his brief demonstration has concluded, "It seems to be SVO, a subject-verb-object language, that's not terribly shocking." He continues, "There doesn't seem to be any plural marking on the nouns, unless it's tones and I missed it. . . . There's

clearly pitch going on here; whether it's tone remains to be analyzed." From this jargon we can see that when Everett evokes a
word or phrase from his interlocutor, he isn't just parroting back
the sounds; he's drawing a map with theories and hypotheses
about how the language works grounded on years of experience
learning languages.

In addition to his enormous wealth of knowledge as a linguist,
Everett has another trick that gives him an enormous advantage.
The demonstration he has presented is not his own invention.
Called a "monolingual fieldwork" demonstration, this method was
first developed by Everett's teacher Kenneth Pike as a means of
learning indigenous languages. The method lays out a sequence
of objects and actions that the practitioner can use to start piecing
together the language. This method even received some Hollywood exposure after Louise Banks, a fictional linguist, used it to
decode an alien language in the 2016 science fiction movie *Arrival*.

These two pieces in Everett's linguistic arsenal—a richly detailed map of how languages work and a method that provides a
path to fluency—have allowed Everett to accomplish a lot more
than just learning some simple sentences. Over the last thirty
years he has become one of only a handful of outsiders to become
fluent in Pirahã, one of the most unusual and difficult languages
on the planet, spoken only by a remote tribe in the Amazon jungle.

THE POWER OF YOUR METALEARNING MAP

Everett's case beautifully illustrates the power of using metalearning to learn new things faster and more effectively. Being
able to see how a subject works, what kinds of skills and information must be mastered, and what methods are available to do

so more effectively is at the heart of success of all ultralearning projects. Metalearning thus forms the map, showing you how to get to your destination without getting lost.

To see why metalearning is so important, consider one study on the helpful effects of already knowing a second language when learning a third. The study took place in Texas, where monolingual English speakers and bilingual Spanish/English speakers were enrolled in a French class. Follow-up on subsequent tests showed that the bilingual speakers outperformed the monolingual students when learning a new language. On its own, this isn't terribly surprising. French and Spanish are both Romance languages, so there are shared features of grammar and vocabulary that aren't present in English that could conceivably provide an advantage. More interesting, however, is that even among the Spanish/English bilinguals, those who also took Spanish classes ended up doing better when they later needed to learn French. The reason seems to be that taking classes assists with helping form what the study authors call *metalinguistic awareness* in a way that simply knowing a language informally does not. The difference between the two types of bilingual speakers mostly came down to metalearning: one group had content knowledge of the language, but the group that took classes also had knowledge about how information in a language is structured.[*]

Nor is this idea about metalearning restricted to languages. Linguistic examples are often easier to study because there's a cleaner separation of metalearning and regular learning. This is because the contents of unrelated languages, such as vocabulary and grammar, are often quite different, even if the metalearning

---

[*] For our purposes, the terms *metalinguistic awareness* and *metalearning* are interchangeable. The literature is replete with *meta*-terms (metaknowledge, metacognition, metamemory, meta-metacognition, etc.) that have related usages.

structure is the same. Learning French vocabulary won't help you much with learning Chinese vocabulary, but understanding how vocabulary acquisition works in French will likely also help with learning Chinese. By the time my friend and I had reached the last country in our year of learning languages, the process of immersing ourselves and learning a new language from scratch was practically a routine. The words and grammar of Korean may have been completely new, but the process of learning was already well trodden. Metalearning exists in all subjects, but it can often be harder to examine independently from regular learning.

HOW TO DRAW YOUR MAP

Now that you have some idea what metalearning is and its importance to learning faster, how can you apply this to get an edge in your own learning efforts? There are two main ways: over the short term and over the long term.

Over the short term, you can do research to focus on improving your metalearning before and during a learning project. Ultralearning, owing to its intensity and self-directed nature, has the opportunity for a lot higher variance than normal schooling efforts do. A good ultralearning project, with excellent materials and an awareness of what needs to be learned, has the potential to be completed faster than formal schooling. Language learning through intensive immersion can beat lengthy classes. Aggressively paced coding bootcamps can get participants up to a level where they can compete for jobs much faster than those with a normal undergraduate degree. This is because you can tailor your project to your exact needs and abilities, avoiding the one-size-fits-all approach taken in school. However, there's also a danger of choosing unwisely and ending up much worse off. Metalearning research avoids this problem and helps you seek out points where you might even be able to get a significant advantage over the status quo.

Over the long term, the more ultralearning projects you do, the larger your set of general metalearning skills will be. You'll know what your capacity is for learning, how you can best schedule your time and manage your motivation, and you'll have well-tested strategies for dealing with common problems. As you learn more things, you'll acquire more and more confidence, which will allow you to enjoy the process of learning more with less frustration.

In this chapter, I'm going to devote most of the next section to short-term research strategies, since they will probably benefit you the most. However, this emphasis shouldn't undermine the importance of the long-term effects of metalearning. Ultralearning is a skill, just like riding a bicycle. The more practice you get with it, the more skills and knowledge you'll pick up for how to do it well. This long-term advantage likely outweighs the short-term benefits and is what's easiest to mistake for intelligence or talent when seen in others. My hope is that as you get more practice in ultralearning, you'll start to automatically apply many of those skills to learn faster and more effectively.

DETERMINING WHY, WHAT, AND HOW

I find it useful to break down metalearning research that you do for a specific project into three questions: "Why?," "What?," and "How?" "Why?" refers to understanding your motivation to learn. If you know exactly why you want to learn a skill or subject, you can save a lot of time by focusing your project on exactly what matters most to you. "What?" refers to the knowledge and abilities you'll need to acquire in order to be successful. Breaking things down into concepts, facts, and procedures can enable you to map out what obstacles you'll face and how best to overcome them. "How?" refers to the resources, environment, and methods you'll use when learning. Making careful choices here can make a big difference in your overall effectiveness.

With these three questions in mind, let's take a look at each of them and how you can draw your map.

ANSWERING "WHY?"

The first question to try to answer is why are you learning and what that implies for how you should approach the project. Practically speaking, the projects you take on are going to have one of two broad motivations: instrumental and intrinsic.

Instrumental learning projects are those you're learning with the purpose of achieving a different, nonlearning result. Consider the previously mentioned case of Diana Fehsenfeld, who, after a few decades as a librarian, found that her job was becoming obsolete. Computerized file systems and budget cuts meant she would need to learn new skills to stay relevant. She did some research and decided that the best way to do this would be to get a firmer grasp on statistics and data visualization. In this case, she wasn't learning because of her deep love of statistics and data visualization but because she believed that doing so would benefit her career.

Intrinsic projects are those that you're pursuing for their own sake. If you've always wanted to speak French, even though you're not sure how you'll use it yet, that's an intrinsic project. Intrinsic doesn't mean useless. Learning French might have benefits later when you decide to travel or need to work with a client from France at your job. The difference is that you're learning the subject for its own sake, not as a means to some other outcome.

If you're pursuing a project for mostly instrumental reasons, it's often a good idea to do an additional step of research: determining whether learning the skill or topic in question will actually help you achieve your goal. I've often heard stories of people unhappy with their career progress who decide that attending graduate school is the answer. If only they had an MBA or an MA, employers would take them more seriously, they think, and they'd have the career they desire. So they go off to school for two years, rack up tens of thousands of dollars in student debt, and discover that their newly minted credentials don't actually get them much better job opportunities than before. The fix here is to do your research first. Determine if learning a topic is likely to have the effect you want it to before you get started.

Tactic: The Expert Interview Method
The main way you can do research of this kind is to talk to people who have already achieved what you want to achieve. Let's say you want to become a successful architect and think that mastering design skills might be the best step to take. Before you get started, it would be a good idea to talk to some successful architects to get a sense of whether they think your project will actually help with your intended goal. Though this method can be used for many parts of the research process, I've found it particularly valuable for vetting instrumental projects. If someone

who has already accomplished the goal you want to achieve doesn't think your learning project will help reach it or thinks it's less important than mastering some other skill, that's a good sign that your motivation and the project are misaligned.

Finding such people isn't as hard as it sounds. If your goal is career related, look for people who have the career you want and send them an email. You can find them at your current workplace, conferences, or seminars, or even on social networking websites such as Twitter or LinkedIn. If your goal is related to something else, you can search online in forums dedicated to the subject you want to learn. If you want to learn programming, with the goal of building your own apps, for example, you can find online forums dedicated to programming or app development. Then you just need to look for frequent posters who seem to have the knowledge you're looking for and email them.

Reaching out and setting up a meeting with an expert isn't hard, either, but it's a step many people shy away from. Many people, particularly the introverts among us, recoil at the idea of reaching out to a stranger to ask for advice. They worry that they'll be rejected, ignored, or even yelled at for presuming to take up a person's time. The truth is, however, that this rarely happens. Most experts are more than willing to offer advice and are flattered by the thought that someone wants to learn from their experience. The key is to write a simple, to-the-point email, explaining why you're reaching out to them and asking if they could spare fifteen minutes to answer some simple questions. Make the email concise and nonthreatening. Don't ask for more than fifteen minutes or for ongoing mentorship. Though some experts will be happy to help you in those ways, it's not good form to ask for too much in the first email.

What if the person you want to interview doesn't live in your city? In that case, phone or online calls are great alternatives.* Email can work in a pinch, too, but I've found that text often doesn't translate tone well, and you often miss sensing the way the person feels about your project. Saying it's a "great idea" lukewarmly versus enthusiastically can make a world of difference, yet that nuance is missing if you communicate via text only.

Even if your project is intrinsically motivated, asking "Why?" is still very useful. Most learning plans you might choose to emulate will be based on curriculum designers' ideas of what is important for you to learn. If these aren't perfectly lined up with your own goals, you may end up spending a lot of time learning things that aren't important to you or underemphasizing the things that do matter. For these kinds of projects, it's useful to ask yourself what you're trying to learn because it will help you evaluate different study plans for their fit with your goals.

ANSWERING "WHAT?"

Once you've gotten a handle on why you're learning, you can start looking at how the knowledge in your subject is structured. A good way to do this is to write down on a sheet of paper three columns with the headings "Concepts," "Facts," and "Procedures." Then brainstorm all the things you'll need to learn. It doesn't matter if the list is perfectly complete or accurate at this stage. You can always revise it later. Your goal here is to get a rough first pass. Once you start learning, you can adjust the list if you discover that your categories aren't quite right.

---

* Calls on the phone might also avoid unwanted side effects of face-to-face meetings. Women who have tried this method have told me that occasionally their interviewee misinterpret their desire for learning advice as a date.

Concepts

In the first column, write down anything that needs to be understood. Concepts are ideas that you need to understand in flexible ways in order for them to be useful. Math and physics, for example, are both subjects that lean heavily toward concepts. Some subjects straddle the concept/fact divide, such as law, which has legal principles that need to be understood, as well as details that need to be memorized. In general, if something needs to be understood, not just memorized, I put it into this column instead of the second column for facts.

Facts

In the second column, write down anything that needs to be memorized. Facts are anything that suffices if you can remember them at all. You don't need to understand them too deeply, so long as you can recall them in the right situations. Languages, for instance, are full of facts about vocabulary, pronunciation, and, to a lesser extent, grammar. Even concept-heavy subjects usually have some facts. If you're learning calculus, you will need to deeply understand how derivatives work, but it may be sufficient to memorize some trigonometric identities.

Procedures

In the third column, write down anything that needs to be practiced. Procedures are actions that need to be performed and may not involve much conscious thinking at all. Learning to ride a bicycle, for instance, is almost all procedural and involves essentially no facts or concepts. Many other skills are mostly procedural, while others may have a procedural component yet still have facts to memorize and concepts to understand. Learning new vocabulary in a language requires memorizing facts, but pronunciation requires practice and therefore belongs in this column.

USING THIS ANALYSIS TO DRAW YOUR MAP

Once you've finished your brainstorm, underline the concepts, facts, and procedures that are going to be most challenging. This will give you a good idea what the major learning bottlenecks are going to be and can start you searching for methods and resources to overcome those difficulties. You might recognize that learning medicine requires a lot of memorization, so you may invest in a system such as spaced-repetition software. If you're learning mathematics, you might recognize that deep understanding of certain concepts is going to be the tricky spot and consider spending time explaining those concepts to other people so you really understand them yourself. Knowing what the bottlenecks will be can help you start to think of ways of making your study time more efficient and effective, as well as avoid tools that probably won't be too helpful to your goal.

Often this coarse-grained analysis is enough to move on to the next phase of research. However, with more experience, you can dig deeper. You might look at some of the particular features of the concepts, facts, and procedures you're trying to learn to find methods to master them more effectively. When I started my portrait-drawing challenge, for instance, I knew that success would depend highly on how accurately I could size and place facial features. Most people can't draw realistic faces because if those attributes are off even slightly (such as making a face too wide or the eyes too high), they will instantly look wrong to our sophisticated ability to recognize faces. Therefore, I got the idea of doing lots and lots of sketches and comparing them by overlaying the reference photos. That way I could quickly diagnose what kinds of errors I was making without having to guess. If you can't make these kinds of predictions and come up with these kinds of strategies just yet, don't worry. This is the kind of long-term benefit of metalearning that comes from having done more projects.

ANSWERING "HOW?"

Now that you've answered two questions—why you're learning and what you're learning—it's time to answer the final question: How are you going to learn it?

I suggest following two methods to answer how you'll learn something: Benchmarking and the Emphasize/Exclude Method.

Benchmarking

The way to start any learning project is by finding the common ways in which people learn the skill or subject. This can help you design a default strategy as a starting point.

If I'm trying to learn something that is taught in school, say computer science, neurology, or history, one thing I'll do is look at the curricula used in schools to teach that subject. This could be the syllabus from a single class or, as in the case of my MIT Challenge, the course list for an entire degree. When I wanted to learn more about cognitive science, I found a list of textbooks that the University of San Diego's Cognitive Science doctoral program recommends for incoming students without cognitive science backgrounds. Good resources to consider for this approach are universities (MIT, Harvard, Yale, and Stanford are good examples but far from the only ones). Generally course lists and syllabi are available by looking on their websites aimed at existing students.

If I'm trying to learn a nonacademic subject or a professional skill, I'll probably instead do online searches for people who have previously learned that skill or use the Expert Interview Method to focus on resources available for mastering that subject. An hour spent searching online for almost any skill should turn up courses, articles, and recommendations for how to learn it. Investing the time here can have incredible benefits because the quality of the materials you use can create orders-of-magnitude differences in your effectiveness. Even if you're eager to start learning right away, investing a few hours now can save you dozens or hundreds later on.

The Emphasize/Exclude Method

Once you've found a default curriculum, you can consider making modifications to it. I find this easier to do with skills that have obvious success criteria (say drawing, languages, or music) and for which you can generally make a guess at the relative importance to the subject topics prior to studying them. For conceptual subjects or topics where you may not even understand the meaning of the terms in the syllabus, it's probably better to stick closer to your benchmark until you learn a bit more.

The Emphasize/Exclude Method involves first finding areas of study that align with the goals you identified in the first part of your research. If you're learning French with the idea of going to Paris for two weeks and speaking in shops and restaurants, I would focus a lot more on pronunciation than being able to spell correctly. If you're learning programming solely to make your own app, I'd focus on the inner workings of app development more than theories of computation.

The second part of the Emphasize/Exclude Method is to omit or delay elements of your benchmarked curriculum that don't align with your goals. For example, one common recommendation for learning Mandarin Chinese, advocated by people such as the renowned linguist and Sinologist Victor Mair, is to focus on learning to speak before you try to read characters. This isn't the only route available, but if your main goal is to speak, then this path to fluency might be more effective.

HOW MUCH PLANNING SHOULD YOU DO?

One question you may face is when to stop doing research and just get started. The literature on self-directed learning, as typically practiced, demonstrates that most people fail to do a thorough investigation of possible learning goals, methods, and resources. Instead they opt for whatever method of learning comes up naturally in their environment. This clearly leaves a gap between what is practiced and the efficiency that is possible using the best possible method. However, research can also be a way of procrastinating, particularly if the method of learning is uncomfortable. Just doing a bit more research then becomes a strategy to avoid doing the work of learning. There will always be some uncertainty in your approach, so it's important to find the sweet spot between insufficient research and analysis paralysis. You know when you're procrastinating, so just get started.

The 10 Percent Rule

A good rule of thumb is that you should invest approximately 10 percent of your total expected learning time into research prior to starting. If you expect to spend six months learning, roughly four hours per week, that would be equal to roughly one hundred hours, which suggests that you should spend about ten hours, or two weeks, doing your research. This percentage will decrease a little bit as your project scales up, so if you plan to do five hundred or a thousand hours of learning, I don't think it necessarily demands fifty or a hundred hours of research, but maybe closer to 5 percent of your time. The goal here isn't to exhaust every learning possibility but simply to make sure you haven't latched onto the first possible resource or method without thinking through alternatives. Prior to starting my MIT Challenge, I spent roughly six months, part-time, combing through all the course materials. A good idea is to be aware of the common methods of learning, popular resources, and tools along with their strengths and weaknesses before starting. Long projects provide more opportunities for getting derailed and delayed, so doing proper research in the beginning can easily save a much larger amount of time later on.

Diminishing Returns and Marginal Benefit Calculation

Metalearning research isn't a onetime activity you do only before starting your project. You should continue to do research as you learn more. Often obstacles and opportunities aren't clear before you start, so reassessing is a necessary step of the learning process. During my portrait-drawing challenge, for instance, I discovered about halfway that I was getting diminishing returns from my sketch-and-compare method. I realized that I needed a better technique for drawing that had higher accuracy. That led me to do a second round of research, leading to a course taught by Vitruvian Studio, which detailed a more systematic method that greatly increased my accuracy. I hadn't noticed it in my original research because I wasn't aware of the deficiency of my self-developed technique.

A more sophisticated answer to the question of when and how to do research would be to compare the marginal benefits of metalearning to regular learning. One way to do this is to spend a few hours doing more research—interviewing more experts, searching online for more resources, searching for new possible techniques—and then spend a few hours doing more learning along your chosen path. After spending some time on each, do a quick assessment of the relative value of the two activities. If you feel as though the metalearning research contributed more than the hours spent on learning itself, you are likely at a point where more research is still beneficial. If you felt that the extra research wasn't too helpful, you're probably better off sticking to the plan you had before. This type of analysis depends on something known as the Law of Diminishing Returns. This states that the more time you invest in an activity (such as more research), the weaker and weaker the benefits will be as you get closer and closer to the ideal approach. If you keep doing research, eventually it will be less valuable than simply doing more learning, so at that point you can safely focus on learning. In practice, the return to research tends to be lumpy and variable. You might spend a few hours and get nothing, then stumble onto the perfect resource for accelerating your progress. As you finish more projects, it's easier to judge this point intuitively, but the Law of Diminishing Returns and the 10 Percent Rule can provide good approximations for how much research to do and when.

LONG-TERM PROSPECTS FOR METALEARNING

So far we've talked only about the short-term benefits. However, the real benefits of metalearning aren't short term but long term.

They don't reside in a particular project but influence your over-all strengths as a learner.

Each project you do will improve your general metalearning. Every project has the opportunity to teach you new learning methods, new ways to gather resources, better time manage-ment, and improved skills for managing your motivation. Suc-cess in one project will give you confidence to execute your next one with boldness and without self-doubt and procrastination. Ultimately, this effect far outweighs the effect of doing a specific project. Unfortunately, it's also something that can't be boiled down to a tactic or tool. Long-term metalearning is just some-thing you acquire with experience.

The benefits of ultralearning aren't always apparent from the first project because that first project occurs when you're at your lowest level of metalearning ability. Each project you complete will give you new tools to tackle the next, starting a virtuous cycle. Many of the ultralearners I interviewed for this book told me a similar story: that they were proud of their accomplish-ments in individual projects but that the real benefit had been that they now understood the process of learning hard things. That gave them the confidence to pursue other ambitious goals that they wouldn't have even considered previously. This con-fidence and ability are the ultimate goals of ultralearning, even though they're often hard to see from the outset. These benefits, however, can be achieved only by putting in the work. The best research, resources, and strategies are useless unless you follow up with concentrated efforts to learn. That brings us to the next principle of ultralearning: focus.

CHAPTER V

PRINCIPLE 2

Focus

Sharpen Your Knife

> Now I will have less distraction.
> —*Leonhard Euler, mathematician, upon losing the sight in his right eye*

If ever there were an unlikely candidate for scientific greatness, it would have been Mary Somerville. She was born into a poor Scottish family in the eighteenth century, when higher education was not seen as proper for a lady. Her mother did not prevent her from reading, but society at large did not approve of it. An aunt, seeing that behavior, commented to her mother, "I wonder you let Mary waste her time in reading, she never sews more than if she were a man." When she did have an opportunity to attend school briefly, her mother regretted the expense. Somerville explained, "she would have been contented if I had only learnt to write well and keep accounts, which was all that a woman was expected to know." As a woman, she faced even larger obstacles, with household duties and expectations taking precedence over any kind of self-education. "A man can always command his time under the plea of business, a woman is not allowed any such excuse," she lamented. Her first husband, Samuel Greig, was strongly against learning in women.

Yet despite those obstacles, Somerville's accomplishments were vast. She won awards in mathematics, learned several languages to fluency, and knew how to paint and play the piano. In 1835, she, along with the German astronomer Caroline Herschel, were the first women elected to the Royal Astronomical Society. The accomplishment that eventually brought her fame was her translation and expansion of the first two volumes of Pierre-Simon Laplace's *Traité de mécanique céleste*, a massive five-volume work on the theory of gravitation and advanced mathematics, acclaimed as the greatest intellectual achievement since Isaac Newton wrote the *Principia Mathematica*. Laplace himself commented that Somerville was the only woman in the world who understood his work.

The easiest explanation for the vast discrepancy in Somerville's situation and her accomplishments would be genius. It is no doubt true that she possessed an incredibly sharp mind. Her daughter once commented that while she was being taught, her mother could grow impatient. "I well remember her slender white hand pointing impatiently to the book or slate—'Don't you see it? There is no difficulty in it, it is quite clear.'" However, in reading through her descriptions of her life, this seeming genius was beset by many insecurities. She claimed to have "bad memory," recounted struggles learning new things as a child, and had even at one point "thought [herself] too old to learn to speak a foreign language." Whether that was polite modesty or genuine feelings of inadequacy, we cannot know, but it does at least put cracks in the idea that she approached learning from a place of unshakable confidence and talent.

Peering deeper, another picture of Somerville emerges. She had a keen intellect, yes, but what she possessed in even greater quantities was an exceptional ability to focus. As an adolescent, when she was put to bed and denied a candle for reading, she would mentally work through the works of Euclid in mathematics. While still breastfeeding her child, an acquaintance encouraged her to study botany, so she devoted "an hour of study to that science" every morning. Even during her greatest achievement, the translation and expansion of Laplace's *Traité de mécanique céleste*, she had to carry out all the household duties of raising children, cooking, and cleaning. "I was always supposed to be at home," she explains, "and my friends and acquaintances came so far out of their way on purpose to see me, it would have been unkind and ungenerous not to receive them. Nevertheless, I was sometimes annoyed when in the midst of a difficult problem one would enter and say, 'I have come to spend a few hours with you.' However, I learnt by habit to leave a subject and resume it again at once, like putting a mark into a book I might be reading."

In the realm of great intellectual accomplishments an ability to focus quickly and deeply is nearly ubiquitous. Albert Einstein focused so intensely during his formulation of the general theory of relativity that he developed stomach problems. The mathematician Paul Erdős was a heavy user of amphetamines to increase his capacity for focus. When a friend bet him that he could not give them up, even for a short time, he did manage to do so. Later, however, he complained that the only result had been that mathematics as a whole was set back a month in his unfocused absence. In these annals of extreme focus, one often conjures up an image of solitary geniuses laboring away without distraction, free from worldly concerns. However remarkable this is, I'm more interested in the kind of focus that Somerville seemed to possess. How can one in an environment such as hers, with constant distractions, little social support, and continuous obligations, manage to focus long enough not only to learn an impressive breadth of subjects, but to such depths that the French mathematician Siméon Poisson once remarked that "there were not twenty men in France who could read [her] book"?

How did Somerville become so good at focusing? What can we glean from her strategies in getting difficult mental work done in less-than-ideal conditions? The struggles with focus that people have generally come in three broad varieties: starting, sustaining, and optimizing the quality of one's focus. Ultralearners are relentless in coming up with solutions to handle these three problems, which form the basis of an ability to focus well and learn deeply.

PROBLEM 1: FAILING TO START FOCUSING (AKA PROCRASTINATING)

The first problem that many people have is starting to focus. The most obvious way this manifests itself is when you procrastinate: instead of doing the thing you're supposed to, you work on something else or slack off. For some people, procrastination is the constant state of their lives, running away from one task to another until deadlines force them to focus and then having to struggle to get the job done on time. Other people struggle with more acute forms of procrastination that manifest themselves with particular kinds of tasks. I was more like this second kind of person, where there were certain types of activities I would spend all day procrastinating on. Though I have no problems writing essays for my blog, when I had to do research for this book, I dragged my feet. Similarly, I had no problem sitting and watching the videos of MIT classes, but I always tackled the first problem sets with considerable trepidation. Had it not been for the intense schedule I was on, I might have found excuses to avoid doing so for much longer. In fact, writing this chapter was one of the tasks I procrastinated on a great deal.

Why do we procrastinate? The simple answer is that at some level there's a craving that drives you to do something else, there's an aversion to doing the task itself, or both. In my case, I procrastinated on writing this chapter because I had a lot of ideas and I was unsure where to start. My anxiety was that by committing something to paper, there was a good chance I might end up writing it poorly. Silly, I know. But most motives to procrastinate are silly when you verbalize them, yet that doesn't stop them from ruling your life. Which brings me to the first step to overcoming procrastination: recognize when you are procrastinating.

Much procrastination is unconscious. You're procrastinating, but you don't internalize it that way. Instead you're "taking a much-needed break" or "having fun, because life can't always be about work all the time." The problem isn't those beliefs. The problem is when they're used to cover up the actual behavior— you don't want to do the thing you need to be focusing on, either because you are directly averse to doing it or because there's something else you want to do more. Recognizing that you're procrastinating is the first step to avoiding it.

Make a mental habit of every time you procrastinate; try to recognize that you are feeling some desire not to do that task or a stronger desire to do something else. You might even want to ask yourself which feeling is more powerful in that moment—is the problem more that you have a strong urge to do a different activity (e.g., eat something, check your phone, take a nap) or that you have a strong urge to avoid the thing you should be doing because you imagine it will be uncomfortable, painful, or frustrating? This awareness is necessary for progress to be made, so if you feel as though procrastination is a weakness of yours, make building this awareness your first priority before you try to fix the problem.

Once you can easily and automatically recognize your tendency to procrastinate, when it occurs, you can take steps to resist the impulse. One way is to think in terms of a series of "crutches" or mental tools that can help you get through some of the worst parts of your tendency to procrastinate. As you get better about taking action on the project you're working on, these crutches can be changed or gotten rid of altogether when procrastination is no longer a problem.

A first crutch comes from recognizing that most of what is unpleasant in a task (if you are averse to it) or what is pleasant about an alternative task (if you're drawn to distraction) is an impulse that doesn't actually last that long. If you actually start working or ignore a potent distractor, it usually only takes a couple minutes until the worry starts to dissolve, even for fairly unpleasant tasks. Therefore, a good first crutch is to convince yourself to get over just the few minutes of maximal unpleasantness before you take a break. Telling yourself that you need to spend only five minutes on the task before you can stop and do something else is often enough to get you started. After all, almost anyone can endure five minutes of anything, no matter how boring, frustrating, or difficult it may be. However, once you start, you may end up continuing for longer without wanting to take the break.

As you progress, your first crutch may start to get in the way. You may find yourself starting but then, because the task is unpleasant and focus is hard, taking advantage of the five-minute rule too often to be productive. If this is the case and your problem has switched from being unable to get started to taking breaks too often, you can try something a little harder, say the Pomodoro Technique: twenty-five minutes of focus followed by a five-minute break.* Keep in mind that it's essential not to switch to a harder goal when you're still mostly impeded by an earlier problem. If you still can't start working, even with the five-minute rule, switching to harder and more demanding crutches may backfire.

In some cases, the moment of frustration may not come at the beginning, but still be predictable. When I was learning Chinese characters through flash cards, for instance, I'd always feel an urge to give up whenever I couldn't remember the answer to one of my cards. I knew this feeling was temporary, however, so I added a rule for myself: I can only quit when I've remembered the most recent card correctly. In practice, the cards were quick, so this usually only took an extra twenty or thirty seconds of persistence; however, my patience for doing flash cards went up dramatically as a result.

Eventually, if working on your project is not troubled by extreme procrastination, you may want to switch to using a calendar on which you carve out specific hours of your day in advance to work on the project. This approach allows you to make the best use of your limited time. However, it works only if you actually follow it. If you find yourself setting a daily schedule with

* This time management method comes from an Italian management consultant, Francesco Cirillo. It is so named because *pomodoro* is Italian for "tomato," and the timer he used was shaped like a tomato.

chunked hours and then frequently ignore it to do something else, go back to the start and try building back up again with the five-minute rule and then the Pomodoro Technique.

Eventually, you may reach Mary Somerville's level of focus, one that she could activate on a moment-to-moment basis, making a decision as to whether she had time to spare. Despite her formidable capacity for focus, it seems that even Somerville would deliberately block out time for the study of particular subjects. Therefore it was a conscious habit, not merely spontaneous studying, that enabled her many successes. For myself, I find that some learning activities are so intrinsically interesting that I can focus on them for a long time without pressure. I generally had no problem watching lectures during the MIT Challenge, for instance. Other tasks, however, required the five-minute rule for me to get past my desire to procrastinate. If I had to scan and upload my files, they'd often build up in a pile before I would finally tackle them. Don't ever feel bad if you have to back up a stage, either; you cannot control your aversions or tendency to distraction, but with practice you can lessen their impact.

PROBLEM 2: FAILING TO SUSTAIN FOCUS (AKA GETTING DISTRACTED)

The second problem people tend to encounter is an inability to sustain focus. This can happen when you've sat yourself down to study or practice something, but then your phone buzzes and you look away, a friend knocks on the door to say hello, or you spin off into a daydream only to realize you've been staring at the same paragraph for the last fifteen minutes. Like the challenge of initiating focus, sustaining focus is important if you want to make progress learning hard things. Before I talk about how to sustain focus, however, I'd like to raise a question about what kind of focus is the best to sustain.

Flow, a concept pioneered by the psychologist Mihály Csíkszentmihályi, is often used as the model for what ideal focus looks like. This is the state of mind you associate with being "in the zone." You stop being interrupted by distracting thoughts, and your mind becomes completely absorbed in the task at hand. Flow is the enjoyable state that slides right between boredom and frustration, when a task is neither too hard nor too easy. This rosy picture, however, does have some detractors. The psychologist K. Anders Ericsson, the researcher behind deliberate practice, argues that flow has characteristics that are "inconsistent with the demands of deliberate practice for monitoring explicit goals and feedback and opportunities for error correction. Hence, skilled performers may enjoy and seek out flow experiences as part of their domain-related activities, but such experiences would not occur during deliberate practice." Ultralearning, with its similar focus on performance-driven learning, would also appear to be unsuitable for flow, in the same way that Ericsson originally argued for deliberate practice.

My own thought is that a flow state is not impossible during ultralearning. Many cognitive activities associated with learning are in the range of difficulty that makes flow possible or even likely. However, I also agree with Ericsson that learning often involves entering into situations in which the difficulty makes flow impossible. Additionally, the self-consciousness that is absent in flow may need to be present in both ultralearning and deliberate practice, as you need to consciously adjust your approach. Working on a programming problem at the limit of your abilities, pushing yourself to write in a style that is unfamiliar to you, or trying to minimize your accent when speaking a new language is each a task that goes against the automatic patterns you may have accumulated. This resistance to what is natural may make flow harder to achieve, even though it is ultimately beneficial for accomplishing your learning goal.

My advice? Don't worry about flow. In some learning tasks, you'll achieve it easily. I often felt as though I were in a flow state while doing practice problems during the MIT Challenge, drilling vocabulary while learning languages, or drawing. At the same time, don't feel guilty if flow doesn't come automatically. Your goal is to enhance your learning, and this often involves pushing through some sessions that are more frustrating than what could be considered ideal for flow. Remember, even if your learning is intense, your use of the skill later on will not be. Investments made in pushing through learning now will make skillful practice a much more enjoyable activity down the road.

After considering how you should focus, let's consider duration. How long should you study? While this problem presumes that you're getting distracted and giving up focusing long before you should, the literature on focus does not suggest that ever-longer periods of focus are optimal from a learning standpoint. Researchers generally find that people retain more of what they learn when practice is broken into different studying periods than when it is crammed together. Similarly, the phenomenon of interleaving suggests that even within a solid block of focus, it can make sense to alternate between different aspects of the skill or knowledge to be remembered. Therefore, if you have several hours to study, you're possibly better off covering a few topics rather than focusing exclusively on one. Doing so has trade-offs, however, so if your study time becomes more and more fractured, it may be difficult to learn at all.

What's needed is a proper balance. To achieve it, fifty minutes to an hour is a good length of time for many learning tasks. If your schedule permits only more concentrated chunks of time, say once per week for several hours, you may want to take several minutes as a break at the end of each hour and split your time over different aspects of the subject you want to learn. Of course, these are merely efficiency guidelines; you ultimately need to find what works best for you, considering not only what is optimal for the purposes of retention but also what fits your schedule, personality, and work flow. For some people, as little as twenty minutes might fit their lives best; others may prefer to spend an entire day learning.

Supposing that you've found a chunk of time to learn that is as optimal for you as it can be, how can you sustain your focus during that time? I've found that there are three different sources that cause focus to break down and distraction to occur. If you're struggling to concentrate, look at each of these three in turn.

Distraction Source 1: Your Environment

The first source of distraction is your environment. Do you have your phone turned off? Are you accessing the internet, watching television, or playing games? Are there distracting noises and sounds? Are you prepared to work, or might you need to stop to look for pens, a book, or a lamp? This is a source of the problem of sustaining focus, but it's also an aspect people frequently ignore for the same reasons they ignore the fact that they are procrastinating. Many people tell themselves that they focus better while listening to music, let's say, but the reality might be that they don't want to work on a given task, so music provides a low-level, amusing distraction. This isn't to condemn anyone who doesn't work in a perfect environment. I certainly don't. Rather, be aware of what environment you work best in, and test it. Do you actually get more work done with the television on in the background, or do you just like hearing the television and feel that it makes the work more bearable? If it's the latter, you can probably train yourself to avoid multitasking and enjoy greater productivity. Multitasking may feel like fun, but it's unsuitable for ultralearning, which requires concentrating your full mind on the task at hand. It's better to rid yourself of this vice than to strengthen bad habits of ineffective learning.

Distraction Source 2: Your Task
The second source is the task you're trying to learn. Certain activities, due to their nature, are harder to focus on than others. I find reading harder to focus on than watching a video, even when the content is the same. Whenever you have a choice between using different tools for learning, you may want to consider which is easier to focus on when making that decision. This choice of materials shouldn't supersede other considerations—I wouldn't opt for a tool that is much less direct (Principle 3) or offers no feedback (Principle 6), simply for the sake of greater focus. Fortunately, these principles are generally aligned, and it is actually the somewhat less effective methods that are less cognitively demanding and therefore harder to sustain focus on. Sometimes you can subtly modify what you're doing to enable greater focus. If I have difficult reading to do, I will often make an effort to jot down notes that reexplain hard concepts for me. I do this mostly because, while I'm writing, I'm less likely to enter into the state of reading hypnosis where I'm pantomiming the act of reading while my mind is actually elsewhere. More intense strategies, whether solving problems, making something,

or writing and explaining ideas aloud, are harder to do in the background of your mind, so there are fewer opportunities for distractions to creep in.

Distraction Source 3: Your Mind

The third source is your mind itself. Negative emotions, restlessness, and daydreaming can be some of the biggest obstacles to focus. This problem has two sides. First, it's obvious that a clear, calm mind is best for focusing on almost all learning problems. A mind filled with angers, anxieties, frustrations, or sadness will be harder to study with. This means that if you're struggling with problems in your life, you'll have a harder time learning well, and you may want to look at dealing with those first. Being in a toxic relationship, having anxiety about some other task you're procrastinating on, or simply knowing you're going down the wrong road in life can interfere with your motivation, so it's often best not to ignore these issues. However, sometimes there's nothing you can do about your emotions, and feelings arise spontaneously without requiring you to do something about them. A random worry about some future event might bubble up, let's say, but you know you shouldn't stop the activity you're working on right now in order to deal with it. Here the solution is to acknowledge the feeling, be aware of it, and gently adjust your focus back to your task and allow the feeling to pass.

Allowing negative feelings to pass, of course, is a lot easier said than done. Emotions can hijack the mind and make the process of returning awareness to your project feel like a Sisyphean task. If I'm really anxious about something, for instance, I may feel as though I'm returning my attention to a task, only for it to jump away fifteen seconds later, repeating again and again for an hour or more. In such moments, recognize that by not reacting to the emotion at the level of abandoning your task entirely, you'll diminish its intensity in the future. You'll also strengthen your commitment to continue working in future situations like this, so they will become easier. Mindfulness researcher and psychiatrist Susan Smalley and meditation teacher Diana Winston of UCLA's Mindful Awareness Research Center argue that when we are engaging in a behavior, our typical reaction is to try to suppress distracting thoughts. If instead you "learn to let it arise, note it, and release it or let it go," this can diminish the behavior you're trying to avoid. If it ever feels as though continuing working is pointless because you're so distracted by a negative emotion that you can't possibly work, remember that the long-term strengthening of your ability to persist on this task will be useful, so the time is not wasted even if you don't accomplish much in this particular learning session.

PROBLEM 3: FAILING TO CREATE THE RIGHT KIND OF FOCUS

A third, problem, subtler than the other two, has to do with the quality and direction of your attention. Supposing you've managed to wrangle the problems of procrastination and distraction down long enough to focus on your task, how should you do it? What's the optimal degree of alertness to maximize your learning?

Here there is some interesting research relating two different variables, arousal and task complexity, to the question of how you should apply your attention. Arousal (the general, not sexual, variety) is your overall feeling of energy or alertness. When you're sleepy, you have low arousal; when you're exercising, you have high arousal. This bodily phenomenon occurs due to sympathetic nervous system activation, and it consists of a range of effects in the body that often occur together, including faster heart rate, increased blood pressure, pupil dilation, and sweating. Mentally, arousal also influences attention. High arousal creates a feeling of keen alertness, which is often characterized by a fairly narrow range of focus, but one that can also be somewhat brittle. This can be very good for focusing on relatively simple tasks or ones that require intense concentration toward a small target. Athletes require this kind of concentration to throw a dart at a target or shoot a basketball properly, where the task is fairly simple but requires concentration to execute properly. Too much arousal, however, and focus starts to suffer. It becomes very easy to be distracted, and you may have a hard time holding focus at any particular spot. Anyone who's drunk too much coffee and feels jittery knows how this can impact your work.

More complex tasks, such as solving math problems or writing essays, tend to benefit from a more relaxed kind of focus. Here the space of focus is often larger and more diffuse. This has advantages when, in order to solve the problem you're facing, you must consider many different inputs or ideas. Trying to solve a complex math problem or write a love sonnet is likely to require this mental quietness. When doing a particularly creative task, if you get stuck, you may benefit from no focus at all. Taking a break from the problem can widen the space of focus enough that possibilities that were not in your consciousness earlier can conjoin and you can make new discoveries. This is a scientific explanation of "Eureka!" moments occurring during leisure or while falling asleep, instead of while at work. Still, before you begin to think that sloth is the key to creativity, it's clear that such an approach often only works when one has been focusing on a problem for long enough that the residue of ideas remains in one's mind. Not working at all is unlikely to lead to creative genius, but taking a break may help breathe fresh perspective into a hard problem.

The relationship between task complexity and arousal is interesting because the latter can be modified. In one experiment, sleep-deprived and well-rested subjects worked on a cognitive task. Unsurprisingly, the sleepy subjects didn't do as well. More interesting, however, was that the sleepy subjects did better when a loud noise was played in the background, while the well-rested subjects did worse. The conclusion drawn by the researchers was that the noise increased arousal levels, which benefited the low-arousal sleepy subjects, but it increased arousal too much for the well-rested ones, causing their decline in performance. This implies that you may want to consider optimizing your arousal levels to sustain the ideal level of focus. Complex tasks may benefit from lower arousal, so working in a quiet room at home might be the right idea for math problems. Simpler tasks might benefit from a noisier environment, say working at a coffee shop. This laboratory experiment shows that you should find out what works best for your own ability to focus through self-testing. You may find you can work better on complex tasks even in the noisy coffee place, or you may find that even for simple tasks you need the quiet room in the library.

IMPROVING YOUR ABILITY TO FOCUS

Focus doesn't need to be exclusive to the domain of those who have endless hours and large swaths of free time on their schedule. As was the case with Somerville, the ability to focus is even more important for those whose lives make such large commitments of time impossible. With practice you can improve your ability to focus. I'm agnostic about whether focus can be trained as an ability, in general. Just because you're disciplined about one thing doesn't automatically make you disciplined about everything else. However, what does generalize is that there is a procedure you can follow to get better at focusing. My advice is this: recognize where you are, and start small. If you're the kind of person who can't sit still for a minute, try sitting still for half a minute. Half a minute soon becomes one minute, then two. Over time, the frustrations you feel learning a particular subject may become transmuted into genuine interest. The impulse to engage in distractions will weaken each time you resist it. With patience and persistence, your few minutes may become large enough to accomplish great things, just as Somerville did almost two hundred years ago.

Now that we've discussed how to get started on learning hard things, let's move to discussing the right way to learn them. The next principle, directness, is the first to explain what types of things you should do while learning and, more important, which you should avoid if you want to be able to use what you learn.

CHAPTER VI

PRINCIPLE 3

Directness

Go Straight Ahead

> He who can go to the fountain does not go to
> the water jar.
> —*Leonardo da Vinci*

After growing up in India, Vatsal Jaiswal moved to Canada with the dream of becoming an architect. Now, four years later, armed with a newly minted degree and entering into the worst job market since the Great Depression, that dream was beginning to seem very far away. Getting a foothold in architecture can be difficult, even in good economic times. But just a few years out from the market crash of 2007, it was nearly impossible. Firms were laying off even experienced architects. If anyone was hiring, they weren't taking chances on some kid just out of school. Out of his graduating class, almost nobody had found an architecture job yet. Most had given up, taking jobs outside the field, going back for more education, or moving in with parents until the economic storms abated.

Another rejection. Jaiswal leaves the offices of yet another architecture firm, walking back to his sliver of the one-bedroom apartment he shares with two roommates. After hundreds of résumés submitted with no reply, he's moved on to trying a more aggressive tactic of going directly to a firm's offices, pleading to speak with whomever is in charge. Still, after weeks of knocking on doors and making dozens of unsolicited office visits, there's no job offer in sight. He hasn't even gotten a call back for a single interview.

Still, Jaiswal suspected that his struggles could be blamed on more than just the recession. From the snippets of feedback he could pry out of the places he applied to, he sensed that the companies didn't see him as a useful employee. He had studied architecture in school, but his program had focused mostly on design and theory. He had been trained in creative design projects that were isolated from the reality of building codes, construction costs, and tricky software. Because his portfolio of school projects didn't resemble the detailed technical documents the architects worked with, they thought hiring him would involve a lengthy training period, something few firms could currently afford.

Jaiswal needed to come up with a plan. More résumé submissions and office walk-ins weren't going to work. He needed a new portfolio that could prove he had the exact skills firms wanted. He needed to show them that, rather than being a burden, he could get to work straightaway and be a valuable team member from the first day.

To do this, he would need to know more about how architects actually drew plans for buildings—not just the big theories and designs, which he had learned in school, but little details of how they did their drawings, what codes they used to represent different materials, and what the drawings showed and omitted. To do that, he found a job at a large-form print shop, the kind that does printing on the large sheets of paper favored for architectural blueprints. Low paying and low skilled, a job in a print shop wasn't Jaiswal's end goal. Still, it could help him scrape by financially while he prepared his new portfolio. Even better, the print store gave him daily exposure to the blueprints firms were using. That allowed him to absorb countless details about how the drawings were put together.

Next, Jaiswal would need to upgrade his technical skills. From his walk-in visits, he was aware that many of the firms he was applying at were using a complex design software called Revit. If he could master its ins and outs, he thought, he could be immediately useful in the technology-heavy entry-level position he desired. At night, he pushed through online tutorials and taught himself the software.

Finally, he was ready to construct a new portfolio. Combining his new Revit knowledge with the knowledge of architectural drawings he had gained while working at the print shop, he made a new portfolio. Instead of the assorted projects from university, he focused on a single building of his own design: a three-tower residential structure with raised courtyards and a modern aesthetic. The project pushed his skills with software further, forcing him to learn new methods and ideas beyond the basics of his online tutorials and exposure at the print shop. Eventually, after a few months of work, he was ready.

New portfolio in hand, Jaiswal submitted it again, this time to just two architecture firms. To his surprise, they both immediately offered him a job.

THE IMPORTANCE OF BEING DIRECT

Jaiswal's story perfectly illustrates the third principle of ultra-learning: directness. By seeing how architecture was actually being done and learning a set of skills that was closely related to the job position he wanted to perform, he was able to cut through the swaths of recent graduates with unimpressive portfolios.

Directness is the idea of learning being tied closely to the situation or context you want to use it in. In Jaiswal's case, when he wanted to get enough architectural skill that firms would hire him, he opted to build a portfolio using the software those firms used and design in the style those firms practiced. There are many routes to self-education, but most of them aren't very direct. In contrast to Jaiswal, another architect I spoke with aimed to improve his employability by deepening his knowledge of design theories. Though that might have been interesting and fun, it was disconnected from the actual skills he would be using in entry-level work. Just as Jaiswal struggled to get work with his university portfolio, many of us are building the wrong portfolio of skills for the kinds of career and personal achievements we want to create. We want to speak a language but try to learn mostly by playing on fun apps, rather than conversing with actual people. We want to work on collaborative, professional programs but mostly code scripts in isolation. We want to become great speakers, so we buy a book on communication, rather than practice presenting. In all these cases the problem is the same: directly learning the thing we want feels too uncomfortable, boring, or frustrating, so we settle for some book, lecture, or app, hoping it will eventually make us better at the real thing.

Directness is the hallmark of most ultralearning projects.* Roger Craig did his *Jeopardy!* testing on the actual questions from past shows. Eric Barone learned video game art by making art for his video game. Benny Lewis learns to speak languages quickly by following a policy of attempting some back-and-forth dialog from the very first day. What these approaches share is that the learning activities are always done with a connection to the context in which the skills learned will eventually be used.

The opposite of this is the approach so often favored in more traditional classroom-style learning: studying facts, concepts, and skills in a way that is removed from how those things will eventually be applied: mastering formulas before you understand the problem they're trying to solve; memorizing the vocabulary of a language because it's written on a list, not because you want to use it; solving highly idealized problems that you'll never see again after graduation.

Indirect approaches to learning, however, aren't limited to traditional education. Many self-directed learners fall into the trap of indirect learning. Consider Duolingo, currently one of the most popular language-learning applications. On the surface, there's a lot to like about this app. It's colorful and fun and gives you a potent sense of progress. But I suspect that much of the sense of progress is an illusion, at least if your goal is to eventually be able to speak the language. To understand why, consider how Duolingo encourages you to practice. It provides English words and sentences and then asks you to pick words from a word bank to translate them.† The problem is that this is

* Directness, as I'm writing about here, is closely related to the concept of transfer-appropriate processing, from psychological literature.
† In fairness to Duolingo, there are ways of using the app to get more direct forms of practice, but these tend to come only from repeatedly practicing the same lessons on the mobile version of the app.

nothing like actually speaking a language! In real life, you may start by trying to translate an English sentence into the language you want to learn. However, real speaking situations don't present themselves as a multiple choice. Instead, you have to dredge up the actual words from memory or find alternative words if you haven't learned one of the ones you want to use. This is, cognitively speaking, quite a different task from picking out matching translations from a highly limited word bank, and also much more difficult. Benny Lewis's method of speaking from the start may be hard, but it transfers perfectly to the task he eventually wants to become good at: having conversations.

During the MIT Challenge, I recognized that the most important resource for being able to eventually pass the classes wasn't having access to recorded lectures, it was having access to problem sets. Yet, in the years since this project, when I am asked for help by students, they often decry the absence of lecture videos from some classes, only rarely complaining about incomplete or insufficient problem sets. This makes me think that most students view sitting and listening to a lecture as the main way that they learn the material, with doing problems that look substantially similar to those on the final exam as being a superficial check on their knowledge. Though first covering the material is often essential to begin doing practice, the principle of directness asserts that it's actually while doing the thing you want to get good at when much of learning takes place. The exceptions to this rule are rarer than they may first appear, and therefore directness has been a thorny problem in the side of education for over a century.

The easiest way to learn directly is to simply spend a lot of time doing the thing you want to become good at. If you want to learn a language, speak it, as Benny Lewis does. If you want to master making video games, then make them, as Eric Barone does. If you want to pass a test, practice solving the kinds of problems that are likely to appear on it, as I did in my own MIT Challenge. This style of learning by doing won't work for all projects. The "real" situation may be infrequent, difficult, or even impossible to create, and thus learning in a different environment is unavoidable. Roger Craig couldn't practice *Jeopardy!* by being on the show hundreds of times. He knew he had to learn in a different environment and prepare to transfer that knowledge to the show when it came time to do so. In such situations, directness isn't an all-or-nothing feature but something you can gradually increase to improve your performance. Craig's approach to start by learning from actual past *Jeopardy!* questions was a lot more effective than if he had just started learning trivia from random topics. Jaiswal was similarly limited when learning architectural skills, as the places he wanted to work wouldn't hire him. However, he worked around that by training on the same software they used and designing a portfolio that was based on the same types of drawings and renderings that were done in actual practice. The twin challenge of directness is that sometimes the exact situation in which you want to use the skill isn't available for easy practice. Even if you can go straight into learning by doing, this approach is often more intense and uncomfortable than passively watching lecture videos or playing around with a fun app. If you don't pay attention to directness, therefore, it's very easy to slip into lousy learning strategies.

One of the big takeaways of Jaiswal's story might not be the triumph of his self-directed learning project but the failure of his formal education. After all, his difficulties started after he had already spent four years studying architecture intensely at university. Why, then, would such a small project, postgraduation, make such a large difference in his employability? To answer that, I'd like to turn to one of the most stubborn and disturbing problems in educational psychology: the problem of transfer.

TRANSFER: EDUCATION'S DIRTY SECRET

Transfer has been called the "Holy Grail of education." It happens when you learn something in one context, say in a classroom, and are able to use it in another context, say in real life. Although this may sound technical, transfer really embodies something we expect of almost all learning efforts—that we'll be able to use something we study in one situation and apply it to a new situation. Anything less than this is hard to describe as learning at all.

Unfortunately, transfer is also something that, despite more than a century of intense work and research, has largely failed to occur in formal education. The psychologist Robert Haskell has said in his excellent coverage of the vast literature on transfer in learning, "Despite the importance of transfer of learning, research findings over the past nine decades clearly show that as individuals, and as educational institutions, we have failed to achieve transfer of learning on any significant level." He later added, "Without exaggeration, it's an education scandal."

The situation is even more disturbing than it sounds. Haskell pointed out, "We expect that there will be transfer of learning, for example, from a high school course in introductory psychology to a college-level introduction to psychology course. It has been known for years, however, that students who enter college having taken a high school psychology course do no better than students who didn't take psychology in high school. Some students who have taken a psychology course in high school do even worse in the college course." In another study, college graduates were asked questions about economic issues and no difference in performance was found between those who had taken an economics class and those who had not.

Providing multiple examples seems to aid transfer a bit, yet the cognitive science researcher Michelene Chi noted that "in almost all the empirical work to date, on the role of example solutions, a student who has studied examples often cannot solve problems that deviate slightly from the example solution." In his book *The Unschooled Mind: How Children Think and How Schools Should Teach*, the developmental psychologist Howard Gardner pointed to the body of evidence showing that even "students who receive honors grades in college-level physics courses are frequently unable to solve basic problems and questions encountered in a form slightly different from that on which they have been formally instructed and tested." Nor has this failure of transfer been limited to schools. Corporate training also suffers, with the former Times Mirror Training Group chairman John H. Zenger writing "Researchers who rigorously evaluate training have said that demonstrable changes following training are hard to find."

The recognition of the failure of general transfer has a history as long as the study of the problem itself. The first attack on the problem came from the psychologists Edward Thorndike and Robert Woodworth in 1901, with their seminal paper "The Influence of Improvement in One Mental Function upon the Efficiency of Other Functions." In it, they attacked the dominant theory of education at the time, so-called formal discipline theory. This theory suggested that the brain was analogous to a muscle, containing fairly general capacities of memory, attention, and reasoning, and that training those muscles, irrespective of the content, could result in general improvement. This was the predominant theory behind universal instruction in Latin and geometry, on the idea that it would help students think better. Thorndike was able to refute this idea by showing that the ability to transfer was much narrower than most people had assumed.

Although studying Latin has fallen out of favor, many educational pundits are reviving new incarnations of the formal discipline theory by suggesting that everyone learn programming or critical thinking in order to improve their general intelligence. Many popular "brain-training" games also subscribe to this view of the mind, assuming that deep training on one set of cognitive tasks will extend to everyday reasoning. It's been more than one hundred years since the verdict came in, yet the allure of a general transfer procedure still has many searching for the Holy Grail.

Despite all this, the situation isn't without hope. Although empirical work and educational institutions have often failed to demonstrate significant transfer, it is not the case that transfer doesn't exist. Wilbert McKeachie, in reviewing the history of transfer, noted that "Transfer is paradoxical. When we want it, we do not get it. Yet it occurs all the time." Whenever you use an analogy, saying something is like something else, you're transferring knowledge. If you know how to ice skate and later learn to Rollerblade, you're transferring skills. As Haskell pointed out, if transfer were really impossible, we would be unable to function.

So what explains the disconnect? Why have educational institutions struggled to demonstrate significant transfer, if transfer is something we all need to function in the world? Haskell suggests that a major reason is that transfer tends to be harder when our knowledge is more limited. As we develop more knowledge and skill in an area, they become more flexible and easier to apply outside the narrow contexts in which they were learned. However, I'd like to add my own hypothesis as an explanation for the transfer problem: most formal learning is woefully indirect.

OVERCOMING THE PROBLEM OF TRANSFER WITH DIRECTNESS

Directness solves the problem of transfer in two ways. The first and most obvious is that if you learn with a direct connection to the area in which you eventually want to apply the skill, the need for far transfer is significantly reduced. Given a century of research showing the difficulties of transfer along with proposed solutions that have failed to provide lasting results, any student must take seriously the notion that transferring what has been learned between very different contexts and situations will be treacherous. If our learning is, as Haskell suggests, "welded to a place or subject matter," it is a lot better that those situations be close to the ones we actually want to use.

Second, I believe that directness may help with transfer to new situations, beyond its more obvious role in preventing the need for far transfer. Many real-life situations share many subtle details with other real-life situations that they never share with the abstract environment of the classroom or textbook. Learning something new rarely depends just on the mass of easily articulated and codified knowledge present but on the myriad tiny details of how that knowledge interacts with reality. By learning in a real context, one also learns many of the hidden details and skills that are far more likely to transfer to a new real-life situation than from the artificial environment of a classroom. Using a personal example, one of the skills I found most important in the no-English project was being able to quickly use a dictionary or translation app on my phone, so I could fill gaps in my linguistic knowledge in midconversation. However, it's exactly this kind of practical skill that is rarely covered in a language-learning curriculum. While this is a trivial example, real-life situations contain thousands of such pieces of skill and knowledge that are necessary if you're going to apply academically learned subjects in the real world.

Ultimately, it will be for researchers to decide whether the Holy Grail of education will ever be found. In the meantime, as learners, we must accept that initial learning efforts often stick stubbornly to the situations we learn them in. The programmer who learns about an algorithm from a class may have trouble recognizing when to use it in her code. The leader who learns a new management philosophy from a business book may go back to working with the same approach she has always used with her employees. My favorite example, though, has to be when a group of friends invited me to join them at a casino. I asked them if their studies ever kept them from enjoying gambling, and they just looked at me blankly. I thought it was funny because the students were actuaries. Their years spent studying statistics in the classroom should have convinced them that you can't expect to beat the house, yet that connection didn't seem to dawn on them. When we learn new things, therefore, we should always strive to tie them directly to the contexts we want to use them in.

Building knowledge outward from the kernel of a real situation is much better than the traditional strategy of learning something and hoping that we'll be able to shift it into a real context at some undetermined future time.

HOW ULTRALEARNERS AVOID THE PROBLEM OF TRANSFER AND LEARN DIRECTLY

Given the problem of transfer and the importance of learning directly, let's look at some of the ways that this is managed in different ultralearning projects. The simplest way to be direct is to learn by doing. Whenever possible, if you can spend a good portion of your learning time just doing the thing you want to get better at, the problem of directness will likely go away. If this isn't possible, you may need to create an artificial project or environment to test your skills. What matters most here is that the cognitive features of the skill you're trying to master and the way you practice it be substantially similar. Consider again Craig's simulation of *Jeopardy!* games by doing questions from old tests. The fact that he was using actual past questions is more important than whether his program matched the signature blue background color present on the show's display. This is because the background color didn't provide any information that would have changed his responses to the questions. The skill he was practicing wasn't changed much by it. In contrast, if he had taken trivia questions from a different game (say the board game Trivial Pursuit) there might have been differences in how questions are typically asked, the topics they are drawn from, or the difficulty level. Worse, if he had spent all his time reading random Wikipedia articles to learn trivia, he wouldn't have been practicing the fundamental skill of recalling answers based on cryptic *Jeopardy!*-style clues at all.

In other cases, what you're trying to achieve may not be a practical skill. Many of the ultralearners I encountered wanted, as their end goal, to understand a subject particularly well, such as Vishal Maini with machine learning and artificial intelligence. Even my own MIT Challenge was based around gaining a deep understanding of computer science, as opposed to a more practical goal of building an app or video game. Though this may seem like a case where directness no longer matters, that really isn't true. It's simply that the place you want to apply these ideas is less obvious and concrete. In Maini's case, he wanted to be able to think and talk intelligently about machine learning, enough to be able to land a nontechnical role in a company that utilized those methods. That meant that being able to communicate his ideas articulately, understanding the concepts clearly, and being able to discuss them with both knowledgeable practitioners and laypeople was important. That's why his goal to make a mini-course explaining the basics of machine learning fit so well. His learning was directly connected with where he wanted to apply the skill: communicating it to others.

Although the findings of the research on transfer are fairly bleak, there is a glimmer of hope, which is that gaining a deeper knowledge of a subject will make it more flexible for future transfer. Whereas the structures of our knowledge start out brittle, welded to the environments and contexts we learn them in, with more work and time they can become flexible and can be applied more broadly. This is the conclusion of Robert Haskell, and although it does not provide a short-term solution to the problem for new learners, it does suggest a path out for those who want to continue working on a subject until they master it. Many ultralearners who have specialized in a smaller subset of fields are masters at transfer; no doubt this is largely due to their depth of knowledge, which makes transfer easier to accomplish. Dan Everett, who was featured in the opening of the chapter on the first principle, metalearning, is a prime example of this. His linguistic depth allows him to learn new languages relatively easily, compared to someone who has learned only a second language or has learned only languages academically.

HOW TO LEARN DIRECTLY

Given the well-documented difficulty with indirect forms of learning, why are they still the default both in schools and in many failed attempts at self-education? The answer is that learning directly is hard. It is often more frustrating, challenging, and intense than reading a book or sitting through a lecture. But this very difficulty creates a potent source of competitive advantage for any would-be ultralearner. If you're willing to apply tactics that exploit directness despite these difficulties, you will end up learning much more effectively.

Let's examine some of the tactics ultralearners use to maximize this principle and take advantage of the inadequacies of more typical schooling.

Tactic 1: Project-Based Learning
Many ultralearners opt for projects rather than classes to learn the skills they need. The rationale is simple: if you organize your learning around producing something, you're guaranteed to at

least learn how to produce that thing. If you take classes, you may spend a lot of time taking notes and reading but not achieve your goal.

Learning to program by creating your own computer game is a perfect example of project-based learning. Engineering, design, art, musical composition, carpentry, writing, and many other skills naturally lend themselves to projects that produce something at the end. However, an intellectual topic can also be the basis of a project. One ultralearner I interviewed, whose project is still ongoing, wanted to learn military history. His project, in this case, was to work toward producing a thesis paper. Since his end goal was to be able to converse knowledgeably about the subject, a project to produce an original paper applied learning more directly than simply trying to read a lot of books without creating anything.

Tactic 2: Immersive Learning

Immersion is the process of surrounding yourself with the target environment in which the skill is practiced. This has the advantage of requiring much larger amounts of practice than would be typical, as well as exposing you to a fuller range of situations in which the skill applies.

Learning a language is the canonical example of where immersion works. By immersing yourself in an environment where a language is spoken, not only do you guarantee that you'll end up practicing the language a lot more than you would otherwise (since you have no choice), but you also face a broader diversity of situations that require learning new words and phrases. However, language learning is not the only place where you can apply immersion to learn more. Joining communities of people who are actively engaged in learning can have a similar impact, since it encourages constant exposure to new ideas and challenges. For example, novice programmers might join open-source projects to expose themselves to new coding challenges.

Tactic 3: The Flight Simulator Method

Immersion and projects are great, but for many skills there's no way to actually practice the skill directly. For skills such as piloting a plane or performing surgery, it's not even legal to practice them in a real situation until you've already invested considerable time into training. How can you overcome this?

It's important to note that what matters for transfer is not every possible feature of the learning environment, such as what room you're in or what clothes you're wearing while you learn. Rather, it's the cognitive features—situations where you need to make decisions about what to do and cue knowledge you've stored in your head. This suggests that when direct practice is impossible, a simulation of the environment will work to the degree to which it remains faithful to the cognitive elements of the task in question. For flying a plane, this means that practicing on a flight simulator may be as good for learning as flying an actual plane if it sufficiently calls on the discriminations and decisions a pilot needs to make. Better graphics and sounds aren't important, unless they change the nature of the decisions being made or the cues pilots receive for when to use certain skills or knowledge.

When evaluating different methods for learning, those that significantly simulate the direct approach will transfer a lot better. Therefore, if you're trying to evaluate what's the best way to learn French before your trip to France, you'll get more (although not perfect) transfer from doing Skype tutoring than you will from flipping through flash cards.

Tactic 4: The Overkill Approach

The last method I've found for enhancing directness is to increase the challenge, so that the skill level required is wholly contained within the goal that is set. Tristan de Montebello, when preparing to compete in the World Championship of Public Speaking, pushed to speak at middle schools, giving early versions of his talk. His feeling was that the feedback he received at Toastmasters clubs might be too soft or congratulatory to cut deep at what worked and didn't work in his speech. Middle school students, in contrast, would be merciless. If a joke he said wasn't funny or his delivery was boring or cheesy, he would be able to tell immediately from their faces what needed to be reworked. The overkill approach is to put yourself into an environment where the demands are going to be extremely high, so you're unlikely to miss any important lessons or feedback.

Going into this environment can feel intense. You may feel as though you're "not ready" to start speaking a language you've barely learned. You may be afraid to stand onstage and deliver a speech you haven't memorized perfectly. You might not want to dive right into programming your own application and prefer to stick to watching videos where someone else does the coding. But these fears are often only temporary. If you can get enough motivation to start this method, it's often a lot easier to continue it long term. The first week in each new country in my language learning project was always a shock, but soon it became completely normal to live entirely within the new language.

One way you can overkill a project is to aim for a particular test, performance, or challenge that will be above the skill level you strictly require. Benny Lewis likes to attempt language exams, because they provide a concrete challenge. In his German project, he wanted to attempt the highest-level exam, because his awareness of that goal would push him to study more than he might if he were satisfied with comfortable conversations alone. Another friend of mine decided to exhibit her photography as a means of pushing her skills and talent. Deciding in advance that your work will be viewable publicly alters your approach to learning and will gear you toward performance in the desired domain, rather than just checking off boxes of facts learned.

LEARN STRAIGHT FROM THE SOURCE

Learning directly is one of the hallmarks of many of the successful ultralearning projects I've encountered, particularly because of how different it can be from the style of education most of us are used to. Whenever you learn anything new, it's a good habit to ask yourself where and how the knowledge will manifest itself. If you can answer that, you can then ask whether you're doing anything to tie what you're learning to that context. If you're not, you need to tread carefully, as the problem of transfer may rear its ugly head.

The act of learning directly, however, is only half of the answer to the question of what you should do to learn well. Doing a lot of direct practice in the environment where you want to eventually use your skills is an important start. However, in order to master skills quickly, bulk practice isn't enough. This brings us to our next principle of ultralearning: drill.

CHAPTER VII

PRINCIPLE 4

Drill

Attack Your Weakest Point

> Take care of the bars and the piece will take care of itself.
> —*Philip Johnston, composer*

Of all the roles Benjamin Franklin played throughout his life— entrepreneur, inventor, scientist, diplomat, and founding father of the United States—he was first and foremost a writer. It was in writing that he first found success. After fleeing Boston to escape the final years of his indentured labor as an apprentice to his brother's printing company, he went to Philadelphia. There, penniless and unknown, he first worked for another printing company before establishing himself as a competitor. His *Poor Richard's Almanack* became an international bestseller and allowed him to retire at forty-two. However, it was in the latter half of his life that his writing would have world-changing consequences.

As a scientist, Franklin was bad at math and more interested in practical consequences than in grand theories of the universe. However, his prose was "written equally well for the uninitiated as well as the philosopher," noted the English chemist Sir Humphrey Davy, adding "he has rendered his details as amusing as well as perspicuous." The strength of his writing and its practical consequences made him an international sensation.

In politics, it was again Franklin's writing talent that helped him win allies and persuade potential antagonists. Prior to the American Revolution, he penned an essay, supposedly written by King Frederick II of Prussia, entitled "An Edict by the King of Prussia." In it he satirized British-American relations by proposing that, due to early settlers of the British Isles being of German origin, "revenue [should] be raised from said colonies in Britain" by the Prussian king.

Later, his skill with a pen would make his writing into the Declaration of Independence, where he edited Thomas Jefferson's words to become the now famous "We hold these truths to be self-evident."

With such an amazing skill for writing and persuasion, it's worth asking how Franklin acquired it. Fortunately, unlike so many other great writers, whose efforts at honing their skills remain mysterious, we have Franklin's own words for how he did it. In his *Autobiography*, he details his sophisticated efforts to slice apart sections of his writing skill for practice as a young boy. Beginning with a childhood debate against a friend about the merits of educating women (Franklin was for, his friend against), his father noticed that aspects of his writing lacked persuasive ability. Franklin thus "determined to endeavor at improvement" and went about a series of exercises to practice his writing skill.

One such exercise he documents was taking a favorite magazine of his, *The Spectator*, and taking notes on articles that appeared there. He would then leave the notes for a few days and come back to them, trying to reconstruct the original argument from memory. After finishing, he "compared my Spectator with the original, discovered some of my faults, and corrected them." Realizing that his vocabulary was limited, he developed another strategy. By turning the prose into verse, he could replace words with synonyms that matched in meter or rhyme. To improve his sense of the rhetorical flow of an essay, he tried his imitation approach again, but this time he jumbled up the hints so he would have to determine the correct order of the sequence of ideas as he wrote again.

Once he had established some of the mechanics of writing, he moved on to the more difficult task of writing in a style that would persuade. When reading an English grammar book, he was exposed to the idea of the Socratic method, of challenging another's ideas through probing questions rather than direct contradiction. He then went to work, carefully avoiding "abrupt contradiction and positive argumentation," instead focusing on being the "humble inquirer and doubter."

Those early efforts produced results. At age sixteen, he wanted to try to get his work published. Fearing that his elder brother might reject it out of hand, however, he disguised his penmanship and submitted his essay under the pseudonym Silence Dogood, purporting to be a widowed woman living in the countryside. His brother, not knowing the true author, approved and published the essay, so Franklin returned and wrote more. Although initiated as a ruse to have his writing considered fairly, Franklin's practice in adopting other characters would prove invaluable in his later career. *Poor Richard's Almanack*, for example, was written from the perspective of a simple husband and wife, Richard and Bridget

Saunders, and his political essays such as his "An Edict by the King of Prussia" similarly made use of his flexibility to adopt imagined perspectives.

It's difficult to imagine Franklin having become the household name he is today without his having first established a mastery of writing. Whether it was business, science, or statecraft, the unchanging core of what made him persuasive and great was his ability to write well. What distinguished Franklin wasn't merely the amount he wrote or his raw talent but how he practiced. The way in which he decided to break apart the skill of writing and practice its elements in isolation enabled him to master writing at a young age and apply it to the other pursuits for which he would later become famous. Such careful analysis and deliberate practice forms the basis for the fourth ultralearning principle: drill.

THE CHEMISTRY OF LEARNING

In chemistry, there's a useful concept known as the rate-determining step. This occurs when a reaction takes place over multiple steps, with the products of one reaction becoming the reagents for another. The rate-determining step is the slowest part of this chain of reactions, forming a bottleneck that ultimately defines the amount of time needed for the entire reaction to occur. Learning, I'd like to argue, often works similarly, with certain aspects of the learning problem forming a bottleneck that controls the speed at which you can become more proficient overall.

Consider learning mathematics. This is a complex skill that has many different parts: you need to be able to understand the fundamental concepts, you need to be able to remember the algorithm for solving a certain type of problem, and you need to know in what context it applies. Underlying this ability, however, is the ability to do arithmetic and algebra so as to be able to solve the problems in question. If your arithmetic is weak or your algebra sloppy, you'll get the wrong answers even if you've mastered the other concepts.

Another rate-determining step could be vocabulary when learning a foreign language. The number of sentences you can successfully utter depends on how many words you know. If you know too few, you won't be able to talk about very much. If you were able to suddenly inject hundreds of new words into your mental database, you might drastically expand your fluency even if your pronunciation, grammar, or other linguistic knowledge remains unchanged.

This is the strategy behind doing drills. By identifying a rate-determining step in your learning reaction, you can isolate it and work on it specifically. Since it governs the overall competence you have with that skill, by improving at it you will improve faster than if you try to practice every aspect of the skill at once. That was Franklin's insight that allowed him to rapidly improve his writing: by identifying components of the overall skill of writing, figuring out which mattered in his situation, and then coming up with clever ways to emphasize them in his practice, he could get better more quickly than if he had just spent a lot of time writing.

DRILLS AND COGNITIVE LOAD

Rate-determining steps in learning—where one component of a complex skill determines your overall level of performance—are a powerful reason to apply drills. However, they aren't the only one. Even if there isn't one isolatable aspect of the skill that is holding back your performance, it may still be a good idea to apply drills.

The reason is that when you are practicing a complex skill, your cognitive resources (attention, memory, effort, etc.) must be spread over many different aspects of the task. When Franklin was writing, he had to consider not only the logical content of the argument he was making but word choice and rhetorical style. This can create a learning trap. In order to improve your performance in one aspect, you may need to devote so much attention to that one aspect that the other parts of your performance start to go down. If you can judge yourself only on how much you improve at the overall task, it can lead to a situation in which your improvement slows down because you will be getting worse at the overall task while becoming better at a specific component of it.

Drills resolve this problem by simplifying a skill enough that you can focus your cognitive resources on a single aspect. When Franklin focused on reconstructing the order of an essay he had read previously, he could devote all his attention to asking what sequence of ideas leads to a good essay rather than also needing to worry about word usage, grammar, and the content of the arguments.

Astute readers will probably notice a tension between this principle and the last. If direct practice involves working on a whole skill nearest to the situation in which it will eventually be used, drills are a pull in the opposite direction. A drill takes the direct practice and cuts it apart, so that you are practicing only an isolated component. How can you resolve this contradiction?

THE DIRECT-THEN-DRILL APPROACH

The tension between learning directly and doing drills can be resolved when we see them as being alternating stages in a larger cycle of learning. The mistake made in many academic strategies for learning is to ignore the direct context or abstract it away, in the hope that if enough component skills are developed, they will eventually transfer. Ultralearners, in contrast, frequently employ what I'll call the Direct-Then-Drill Approach.

The first step is to try to practice the skill directly. This means figuring out where and how the skill will be used and then trying to match that situation as close as is feasible when practicing. Practice a language by actually speaking it. Learn programming by writing software. Improve your writing skills by penning essays. This initial connection and subsequent feedback loop ensure that the transfer problem won't occur.

The next step is to analyze the direct skill and try to isolate components that are either rate-determining steps in your performance or subskills you find difficult to improve because there are too many other things going on for you to focus on them. From here you can develop drills and practice those components separately until you get better at them.

The final step is to go back to direct practice and integrate what you've learned. This has two purposes. The first is that even in well-designed drills, there are going to be transfer hiccups owing to the fact that what was previously an isolated skill must be moved to a new and more complex context. Think of this as being like building the connective tissue to join the muscles you strengthened separately. The second function of this step is as a check on whether your drill was well designed and appropriate. Many attempts to isolate a drill may end in failure because the drill doesn't really cut at the heart of what was difficult in real practice. That's okay; this feedback is important to help you minimize wasting time learning things that don't matter much to your end goals.

The earlier you are in the learning process, the faster this cycle should be. Cycling between direct practice and drills, even within the same learning session, is a good idea when you're just starting out. Later, as you get better at what you are trying to do and a lot more effort is required to noticeably improve your overall performance, it's more acceptable to take longer detours into drills. As you approach mastery, your time may end up focused mostly on drills as your knowledge of how the complex skill breaks down into individual components becomes more refined and accurate and improving any individual component gets harder and harder.

TACTICS FOR DESIGNING DRILLS

There are three major problems when applying this principle. The first is figuring out when and what to drill. You should focus on what aspects of the skill might be the rate-determining steps in your performance. Which aspect of the skill, if you improved it, would cause the greatest improvement to your abilities overall for the least amount of effort? Your accounting skills might be limited by the fact that your Excel knowledge is superficial, which prevents you from applying all the things you know to practical situations. Your language abilities may be held back by having inaccurate pronunciation, even though you know the right words. Look as well to aspects of a skill that you need to juggle simultaneously. These may be harder to improve because you can't devote enough cognitive resources to improving them. When writing a new article, you may have to juggle research, storytelling, vocabulary, and many other aspects simultaneously, making it hard to get a lot better at just one. Determining what to drill may seem tricky, but it doesn't have to be. The key is to experiment. Make a hypothesis about what is holding you back, attack it with some drills, using the Direct-Then-Drill Approach, and you can quickly get feedback about whether you're right.

The second difficulty with this principle is designing the drill to produce improvement. This is often hard because even if you recognize an aspect of your performance you're weak on, it may be tricky to design a drill that trains that component without artificially removing what makes it difficult in actual application. Franklin's drills were uncommon, I believe, because most people, even recognizing specific deficits in their writing ability, would not have had the ingenuity to find ways to drill subskills such as ordering arguments persuasively and emulating a successful writing style.

Finally, doing drills is hard and often uncomfortable. Teasing out the worst thing about your performance and practicing that in isolation takes guts. It's much more pleasant to spend time focusing on things you're already good at. Given this natural tendency, let's look at some good ways to do drills so you can start applying them yourself.

Drill 1: Time Slicing
The easiest way to create a drill is to isolate a slice in time of a longer sequence of actions. Musicians often do this kind of training when they identify the hardest parts of a piece of music and practice each one until it's perfect before integrating it back into

the context of the entire song or symphony. Athletes similarly engage in this process when they drill skills that are normally a fraction of total playing time, such as layups or penalty shots. In the early phase of learning a new language, I often obsessively repeat a few key phrases, so they quickly get embedded into my long-term memory. Look for parts of the skill you're learning that can be decomposed into specific moments of time that have heightened difficulty or importance.

Drill 2: Cognitive Components

Sometimes what you'll want to practice isn't a slice in time of a larger skill but a particular cognitive component. When speaking a language, grammar, pronunciation, and vocabulary occur at all moments, but they form different cognitive aspects that must be managed simultaneously. The tactic here is to find a way to drill only one component when, in practice, others would be applied at the same time. When learning Mandarin Chinese, I would do tone drills that involved pronouncing pairs of words with different tones and recording myself speaking. That allowed me to practice producing different tones quickly, without the distraction of needing to remember what the words meant or how to form grammatically correct sentences.

Drill 3: The Copycat

A difficulty with drills in many creative skills is that it is often impossible to practice one aspect without also doing the work of the others. When Franklin was trying to improve his ability to order arguments logically, for instance, it wasn't possible to do so without writing an entire essay. To solve this problem in your own learning, you can take a page from Franklin: by copying the parts of the skill you don't want to drill (either from someone else or your past work), you can focus exclusively on the component you want to practice. Not only does this save a lot of time, because you need to repeat only the part you're drilling, it also reduces your cognitive burden, meaning you can apply more focus to getting better at that one aspect. When practicing drawing, I started by drawing not just from photos but from drawings other people had done. That helped me focus on the skill of accurately rendering the picture, simplifying the decision about how to frame the scene and which details to include. For flexible creative works, editing works you've created in the past may have the same effect, allowing you to selectively improve an aspect of your work without having to consider the other demands of an original composition.

Drill 4: The Magnifying Glass Method

Suppose you need to create something new and can't edit or separate out the part you want to practice. How can you create a drill? The Magnifying Glass Method is to spend more time on one component of the skill than you would otherwise. This may reduce your overall performance or increase your input time, but it will allow you to spend a much higher proportion of your time and cognitive resources on the subskill you want to master. I applied this method when trying to improve my ability to do research when writing articles, by spending about ten times as long on research as I had previously. Although I still had to do all the other parts of writing the article, by spending much longer on research than I would normally, I could develop new habits and skills for doing so.

Drill 5: Prerequisite Chaining

One strategy I've seen repeatedly from ultralearners is to start with a skill that they don't have all the prerequisites for. Then, when they inevitably do poorly, they go back a step, learn one of the foundational topics, and repeat the exercise. This practice of starting too hard and learning prerequisites as they are needed can be frustrating, but it saves a lot of time learning subskills that don't actually drive performance much. Eric Barone, for instance, started his pixel art experiments simply by making them. When he struggled with certain aspects, such as colors, he went back, learned color theory, and repeated his work. Benny Lewis has a similar habit of starting with speaking from a phrase book and only later learning the grammar that explains how the phrases function.

MINDFUL DRILLING

To many, the idea of drilling may seem to be a push in the wrong direction. We've all spent time doing homework designed to drill into us facts and procedures that turned out to be a total waste of time. That was often because we didn't know the reasons behind what we were practicing or how it fit into a broader context. Drilling problems without context is mind-numbing. However, once you've identified that it's the bottleneck preventing you from going further, they become instilled with new purpose. In ultralearning, which is directed by the student, not an external source, drills take on a new light. Instead of being forced to do them for unknown purposes, it is now up to you to find a way to enhance the learning process by accelerating learning on the specific things that you find most difficult. In this sense, drills take on a very different flavor in ultralearning as opposed to traditional learning. Far from being meaningless drudgery, carefully designed drills elicit creativity and imagination as you strive to solve a more complex learning challenge by breaking it into specific parts.

Drills are hard to do, which is why many of us would rather avoid them. When we do engage in drills, it's often in subjects where we feel competent and comfortable. Drills require the learner not only to think deeply about what is being learned but also figure out what is most difficult and attack that weakness directly rather than focus on what is the most fun or what has already been mastered. This requires strong motivation and a comfort with learning aggressively. Franklin, in his *Autobiography*, remarked about the lengths he went to so he could dedicate himself to his writing drills: "My time for these exercises and for reading was at night, after work or before it began in the morning." Despite the prominence writing would play in his life, Franklin still had to work long hours under his taskmaster brother in the print shop, diligently improving his craft in what little leisure time he had. Eric Barone similarly repeated his pixel art dozens of times, going back to master prerequisite concepts and theory until he got it perfect.

The difficulty and usefulness of drills repeat a pattern that will recur throughout the ultralearning principles: that something mentally strenuous provides a greater benefit to learning than something easy. Nowhere is this pattern more clear than in the next principle, retrieval, where difficulty itself may be the key to more effective learning.

CHAPTER VIII

PRINCIPLE 5

Retrieval

Test to Learn

It pays better to wait and recollect by an effort from within, than to look at the book again.
—*William James, psychologist*

In the spring of 1913, the mathematician G. H. Hardy received a letter that would forever define the course of his life. Sent by an accounting clerk working for the Port Trust Office of Madras in India, the letter contained a humble note of introduction along with some startling assertions. The author claimed that he had found theorems for problems that the best mathematical minds of the time had yet to solve. What's more, he claimed that he had "no University education" and had derived these results from his own solitary investigations.

Receiving letters from amateur crackpots who claimed to have solutions to famous problems was a common occurrence for someone of Hardy's stature in mathematics, so at first he simply dismissed the letter as being more of the same. Still, flipping through the several pages of notes attached to the letter, the equations wouldn't leave his mind. When he found himself thinking about them hours later, he brought the letter to the attention of his colleague John Littlewood. As the two of them toyed at trying to prove the strange assertions, they found that some of them they were able to prove with great effort, while others remained, in Hardy's words, "scarcely possible to believe." Maybe, Hardy thought, this wasn't a letter from a crackpot but something rather different.

The formulas written were so bizarre and alien that Hardy remarked, "They must be true because, if they were not true, no one would have had the imagination to invent them." What he only vaguely understood that day was that he had just had his first introduction to one of the most brilliant and bizarre mathematicians of all time, Srinivasa Ramanujan.

RAMANUJAN'S GENIUS

Before writing his letter to Hardy, which changed the course of mathematical history, Ramanujan was a poor, pudgy south Indian boy with a special love of equations. More than anything else, he loved math. In fact, his love of math often got him into difficulties. His unwillingness to study other subjects flunked him out of university. Equations were all he cared about. In his spare time and during stretches of unemployment, he would sit for hours on the bench in front of his family home, slate in hand, playing with formulas. Sometimes he would stay up so late that his mother would need to put food into his hand so he would eat.

As he was thousands of miles away from the center of mathematics of his day, access to high-quality textbooks was quite a challenge for Ramanujan. One resource he did encounter and mined extensively was a volume by George Shoobridge Carr called *A Synopsis of Elementary Results in Pure and Applied Mathematics*. Carr himself was hardly a towering figure of mathematical genius. The book, intended as a guide for students, included large lists of various theorems from different fields of mathematics, usually without explanation or proof. However, even without having proofs or explanations available, Carr's book became a powerful resource in the hands of someone smart and obsessed like Ramanujan. For instead of simply copying and memorizing how certain theorems were derived, he had to figure them out for himself.

Though many commentators of the time, including Hardy, argued that Ramanujan's impoverished upbringing and late access to the cutting edge of mathematics likely did irreparable harm to his genius, modern psychological experiments may offer an alternative perspective, for when Ramanujan dealt only with Carr's extensive list of theorems using his own quirky obsession with mathematical formulas, he was unwittingly practicing one of the most powerful methods known to build a deep understanding.

THE TESTING EFFECT

Imagine you're a student preparing for an exam. You have three choices about how you can allocate your limited studying time. First, you can review the material. You can look over your notes and book and study everything until you're sure you'll remember it. Second, you can test yourself. You can keep the book shut and try to remember what was in it. Finally, you can create a concept map. You can write out the main concepts in a diagram, showing how they're organized and related to other items you need to study. If you can pick only one, which one should you choose to do best on the final exam?

This is essentially the question posed by the psychologists Jeffrey Karpicke and Janell Blunt in one study examining students' choice of learning strategy. In the study, students were divided into four groups, each given the same amount of time but told to use different study strategies: reviewing the text a single time, reviewing it repeatedly, free recall, and concept mapping. In each group, students were asked to predict their score on the upcoming test. Those who did repeated reviewing predicted that they'd score the best, followed by the single-study and concept-mapping groups. Those who practiced free recall (trying to remember as much as they could without looking in the book) predicted the worst for their final performance.

The actual results, however, weren't even close. Testing yourself—trying to retrieve information without looking at the text—clearly outperformed all other conditions. On questions based directly on the content of the text, those who practiced free recall remembered almost 50 percent more than the other groups. How could students, who have spent years getting first-hand experience about what matters to learning, be so misguided about what actually produces results?

One might be tempted to argue that this benefit of self-testing is an artifact of the way success is measured. The principle of directness asserts that transfer is difficult. Since self-testing and actual testing are most similar, perhaps it is this similarity that allows this method to work better. Had the method of evaluation differed, it might be reasonable to suspect that review or concept mapping might come out on top. Interestingly, in another experiment, Karpicke and Blunt showed that this wasn't the explanation, either. In this experiment the final test was to produce a concept map. Despite the overwhelming similarity to the evaluation task, free recall still did better than using concept mapping to study.

Another possible explanation for why self-testing works is feedback. When you review something passively, you don't get any feedback about what you know and don't know. Since tests usually come with feedback, that might explain why students who practiced self-testing beat the concept mappers or passive reviewers. Though it is true that feedback is valuable, once again, retrieval doesn't simply reduce down to getting more feedback. In the experiments mentioned, students were asked to do free recall but weren't provided any feedback about items they missed or got wrong. The act of trying to summon up knowledge from memory is a powerful learning tool on its own, beyond its connection to direct practice or feedback.

This new perspective on learning shows how Carr's book, with its lists of proofs without solutions, could have become, in the hands of someone sufficiently motivated to master them, an incredible tool for becoming brilliant at math. Without the answers at hand, Ramanujan was forced to invent his own solutions to the problems, retrieving information from his mind rather than reviewing it in a book.

THE PARADOX OF STUDYING

If retrieval practice—trying to recall facts and concepts from memory—is so much better for learning, why don't students realize it? Why do many prefer to stick to concept mapping or the even less effective passive review, when simply closing the book and trying to recall as much as possible would help them so much more?

Karpicke's research points to a possible explanation: Human beings don't have the ability to know with certainty how well they've learned something. Instead, we need to rely on clues from our experience of studying to give us a feeling about how well we're doing. These so-called judgments of learning (JOLs) are based, in part, on how fluently we can process something. If the learning task feels easy and smooth, we are more likely to believe we've learned it. If the task feels like a struggle, we'll feel we haven't learned it yet. Immediately after spending some time studying, these JOLs may even be accurate. Minutes after studying something using a strategy of passive review, students perform better than they would if they had practiced retrieval. The feeling that you're learning more when you're reading rather than trying to recall with a closed book isn't inaccurate. The problem comes after. Test again days later, and retrieval practice beats passive review by a mile. What helped in the immediate time after studying turns out not to create the long-term memory needed for actual learning to take place.

Another explanation for why students opt for low-efficiency review instead of retrieval is that they don't feel they know the material well enough to test themselves on it. In another experiment, Karpicke had students choose a strategy for learning. Inevitably, students who were performing more weakly elected to review the material first, waiting until they were "ready" to start practice testing. If through experimental intervention, however, they were forced to practice retrieval earlier, they learned more. Whether you are ready or not, retrieval practice works better.

Especially if you combine retrieval with the ability to look up the answers, retrieval practice is a much better form of studying than the ones most students apply.

IS DIFFICULTY DESIRABLE?

What makes practicing retrieval so much better than review? One answer comes from the psychologist R. A. Bjork's concept of desirable difficulty. More difficult retrieval leads to better learning, provided the act of retrieval is itself successful. Free recall tests, in which students need to recall as much as they can remember without prompting, tend to result in better retention than cued recall tests, in which students are given hints about what they need to remember. Cued recall tests, in turn, are better than recognition tests, such as multiple-choice answers, where the correct answer needs to be recognized but not generated. Giving someone a test immediately after they learn something improves retention less than giving them a slight delay, long enough so that answers aren't in mind when they need them. Difficulty, far from being an obstacle to making retrieval work, may be part of the reason it does so.

The idea of desirable difficulties in retrieval makes a potent case for the ultralearning strategy. Low-intensity learning strategies typically involve either less or easier retrieval. Pushing difficulty higher and opting for testing oneself well before you are "ready" is more efficient. One can think back to Benny Lewis's strategy of speaking a new language from the first day. Though this approach is high in difficulty, research suggests why it might be more useful than easier forms of classroom study. Placing himself in a more difficult context means that every time Lewis needs to recall a word or phrase, it will be remembered more strongly than when doing the same act of retrieval in a classroom setting and much better than when simply looking over a list of words and phrases.

Difficulty can become undesirable if it gets so hard that retrieval becomes impossible. Delaying the first test of a newly learned fact has some benefits over testing immediately. However, if you delay the test too long, the information may be forgotten entirely. The idea, therefore, is to find the right midpoint: far enough away to make whatever is retrieved remembered deeply, not so far away that you've forgotten everything. Although waiting too long before you test yourself may have disadvantages, increasing difficulty by giving yourself fewer clues and prompts are likely helpful, provided that you can get some feedback on them later.

SHOULD YOU TAKE THE FINAL EXAM BEFORE THE CLASS EVEN BEGINS?

The standard way of viewing tests is that they work to evaluate the knowledge you have learned elsewhere—through reading or listening to lectures. The concept of retrieval flips this view on its head, suggesting that the act of taking a test not only is a source of learning but results in more learning than a similar amount of time spent in review. However, this still fits within the conventional idea of knowledge being first acquired, and then strengthened or tested later.

An interesting observation from retrieval research, known as the forward-testing effect, shows that retrieval not only helps enhance what you've learned previously but can even help prepare you to learn better. Regular testing of previously studied information can make it easier to learn new information. This means that retrieval works to enhance future learning, even when there is nothing to retrieve yet!

A variety of mechanisms has been proposed for explaining why this forward testing effect exists. Some researchers argue that it may be that trying to find knowledge that hasn't been learned yet—say, by trying to solve a problem you haven't learned the answer to yet—nonetheless helps reinforce search strategies that are put to use once the knowledge is encountered later. An analogy here is that trying to retrieve an answer that doesn't yet exist in your mind is like laying down a road leading to a building that hasn't been constructed yet. The destination doesn't exist, but the path to get to where it will be, once constructed, is developed regardless. Other researchers argue that the mechanism might be one of attention. By confronting a problem you don't yet know how to answer, your mind automatically adjusts its attentional resources to spot information that looks like a solution when you learn it later. Whatever the exact mechanism is, the reality of the forward-testing effect implies that practicing retrieval might not only benefit from starting earlier than one is "ready" but even before you have the possibility of answering correctly.

WHAT SHOULD BE RETRIEVED?

The research is clear: if you need to recall something later, you're best off practicing retrieving it. However, this neglects an important question: What kinds of things should you invest the time in to remember in the first place? Retrieval may take less time than review to get the same learning impact, but not learning something at all is faster still. This is an important practical question. Nobody has time to master everything. During the MIT Challenge, I covered a lot of different ideas. Some were directly relevant to the kind of programming I wanted to do when I was done, so making sure I retained those ideas was a priority. Others were interesting, but since I had no plans to use them immediately, I put more effort into practicing retrieving the underlying concepts than doing technical calculations. One class I did, for instance, was Modal Logic. As I have no plans to be a logician, I can honestly say, eight years later, that I couldn't prove theorems in modal logic today. However, I can tell you what modal logic is for and when it is used, so if a situation arises in which the techniques I learned in that class might be useful, I'd have a much better time spotting it.* There will always be some things you choose to master and others you satisfy yourself with knowing you can look up if you need to.

One way to answer this question is simply to do direct practice. Directness sidesteps this question by forcing you to retrieve the things that come up often in the course of using the skill. If you're learning a language and need to recall a word, you'll practice it. If you never need a word, you won't memorize it. The advantage of this strategy is that it automatically leads you to learn the things with the highest frequency. Things that are rarely used or that are easier to look up than to memorize won't be retrieved. These tend to be the things that don't matter so much.

The problem with relying on direct practice exclusively is that knowledge that isn't in your head can't be used to help you solve problems. For instance, a programmer may realize a need to use

---

* Modal logic is an extension of propositional logic, allowing you to express ideas such as "should," "usually," or "possibly."

a certain function to solve a problem but forgets how to write it out. Needing to look up the syntax might slow her down, but she will still be able to solve the problem. However, if you don't have enough knowledge stored to recognize when you can use a function to solve your problem, no looking up can help you. Consider that over the last twenty years, the amount of knowledge easily accessible from a quick online search has exploded. Nearly any fact or concept is now available on demand to anyone with a smartphone. Yet despite this incredible advance, it is not as if the average person is thousands as times as smart as people were was a generation ago. Being able to look things up is certainly an advantage, but without a certain amount of knowledge inside your head, it doesn't help you solve hard problems.

Direct practice alone can fail to encourage enough retrieval by omitting knowledge that can help you solve a problem but isn't strictly necessary to do so. Consider our programmer who has two different ways to solve her problem, A and B. Option A is much more effective, but B will also get the job done. Now suppose that she knows only about option B. She'll continue to use the way she knows to solve the problem, even though it is less effective. Here, our fledgling programmer might read about option A on a blog somewhere. But since simply reading is much less effective than repeated retrieval practice, chances are that she'll forget about it when it comes time to apply the technique. This may sound abstract, but I'd argue that this is quite common with programmers, and often the thing separating mediocre programmers from great ones isn't the range of problems they can solve but that the latter often know dozens of ways to solve problems and can select the best one for each situation. This kind of breadth requires a certain amount of passive exposure, which in turn benefits from retrieval practice.

HOW TO PRACTICE RETRIEVAL

Retrieval works, but it isn't always easy. Not only is the effort itself an obstacle, but sometimes it's not clear exactly how to do it. Passive review may not be very efficient, but at least it's straightforward: you open your book and reread material until you retain it. Most books and resources don't have a handy list of questions at the end to test you to see if you remember what they contain. To help with that, below are some useful methods that can be used to apply retrieval to almost any subject.

Tactic 1: Flash Cards

Flash cards are an amazingly simple, yet effective, way to learn paired associations between questions and answers. The old way of creating paper flash cards to drill yourself is powerful, but it has largely been superseded by spaced-repetition systems, as I'll discuss in Principle 7. These software algorithms can handle tens of thousands of "cards" and also organize a review schedule so you can manage them.

The major drawback of flash cards is that they work really well for a specific type of retrieval—when there's a pairing between a specific cue and a particular response. For some forms of knowledge, for example memorizing foreign-language vocabulary, this works perfectly. Similarly, maps, anatomical diagrams, definitions, and equations can often be memorized via flash cards. However, when the situation in which you need to remember the information is highly variable, this kind of practice can have drawbacks. Programmers can memorize syntax via flash cards, but concepts that need to be applied in real programs often don't fit the cue-response framework that flash cards demand.

Tactic 2: Free Recall

A simple tactic for applying retrieval is, after reading a section from a book or sitting through a lecture, to try to write down everything you can remember on a blank piece of paper. Free recall like this is often very difficult, and there will be many things missed, even if you just finished reading the text in question. However, this difficulty is also a good reason why this practice is helpful. By forcing yourself to recall the main points and arguments, you'll be able to remember them better later. While doing research for this book, for instance, I would often print out journal articles and put them in a binder with a few blank sheets of paper after each of them. After I had finished reading, I'd do a quick free recall exercise to make sure I would retain the important details when it came time for writing.

Tactic 3: The Question-Book Method

Most students take notes by copying the main points as they encounter them. However, another strategy for taking notes is to rephrase what you've recorded as questions to be answered later. Instead of writing that the Magna Carta was signed in 1215, you could instead write the question "When was the Magna Carta signed?" with a reference to where to find the answer in case you forget. By taking notes as questions instead of answers, you generate the material to practice retrieval on later.

One mistake I've made in applying this technique is to focus on the wrong kinds of things to ask questions about. I tried applying this method to a book on computational neuroscience, and I ended up asking myself all sorts of detailed questions such as what was the firing rate of certain neuronal circuits or who proposed a specific theory. That wasn't intentional but rather a by-product of lazily restating the factual content in the book as questions. What's harder and more useful is to restate the big idea of a chapter or section as a question. Since this is often implicit, it requires some deeper thinking and not just adding a question mark to some notes you copied verbatim. One rule I've found helpful for this is to restrict myself to one question per section of a text, thus forcing myself to acknowledge and rephrase the main point rather than zoom in on a detail that will be largely irrelevant later.

Tactic 4: Self-Generated Challenges

The above tactics work best with retrieval of simple information, such as facts or summaries of broad ideas you might encounter in a book or lecture. However, if you're trying to practice a skill, not merely remember information, they might not be enough. For a programmer, it's not enough to know what an algorithm means, but be able to write it in code. In this case, as you go through your passive material, you can create challenges for yourself to solve later. You may encounter a new technique and then write a note to demonstrate that technique in an actual example. Creating a list of such challenges can serve as a prompt for mastering that information later in practice and can expand your library of tools that you are able to actually apply.

Tactic 5: Closed-Book Learning

Nearly any learning activity can become an opportunity for retrieval if you cut off the ability to search for hints. Concept mapping, the strategy that didn't work particularly well for students in Karpicke and Blunt's experiments, could be beefed up considerably by preventing yourself from looking at the book when generating your concept map. I suspect that had this been done in the original experiment, students using this form of closed-book concept mapping would likely have done better on the eventual test that relied on creating a concept map. Any practice, whether direct or a drill, can be cut off from the ability to look things up. By preventing yourself from consulting the source, the information becomes knowledge stored inside your head instead of inside a reference manual.

REVISITING RAMANUJAN

Ramanujan was smart, there's no denying it. However, his genius was aided immeasurably by two hallmarks of the ultralearner's tool kit: obsessive intensity and retrieval practice. As he worked on his slate from morning to night, trying to figure out Carr's sparsely written list of theorems was incredibly hard work. But it also created the desirable difficulties that allowed him to build a huge mental library of tools and tricks that would assist him in his later mathematical efforts.

Retrieval played an important role in Ramanujan's mathematical upbringing, but he is hardly the only one to take advantage of the tactic. In nearly every biography of great geniuses and contemporary ultralearners I have encountered, some form of retrieval practice is mentioned. Benjamin Franklin practiced his writing by reconstructing essays from memory. Mary Somerville worked through problems mentally when no candle was available for night reading. Roger Craig practiced trivia questions without looking at the answers. Retrieval is not a sufficient tool to create genius, but it may be a necessary one.

Trying to produce the answer rather than merely reviewing it is only half of a bigger cycle, however. To make retrieval really effective, it helps to know whether the answer you dredged up from your mind was correct. Just as we often avoid testing ourselves until we're ready because struggling with a test is uncomfortable, we often avoid seeking information about our skill level until we think it will be favorable. Being able to process that information effectively, hearing the message it contains loud and clear, isn't always easy. Yet this is also why it is so important. This brings us to the next principle of ultralearning: feedback.

CHAPTER IX

PRINCIPLE 6

Feedback

Don't Dodge the Punches

> Everybody has a plan until they get punched
> in the mouth.
> —*Mike Tyson*

From a narrow staircase in the back, Chris Rock enters the stage just as his name is being announced. With sold-out shows and HBO specials, Rock is no neophyte to stand-up comedy. His performances feel like a rock concert. With an energetic and punctuated delivery, he's known for repeating the key phrase of a joke like the chorus of a song, the rhythm of it so precise that you get the feeling he would be able to make anything funny. And that's exactly the problem. When everything you do is funny, how do you know what really makes a joke good?

Far from the packed concert halls and jubilant crowds, Rock

walks to the mic on the modest brick-backed stage at the Comedy Cellar in Greenwich Village, New York City. In his hand are scraps of cards on which he has scribbled bits of phrases, a trick for working out new material he learned from his grandfather, a cab driver who preached on weekends. Instead of his signature aggressive style, he slumps against the back wall. This is his laboratory, and he's going to perform comedy with the precision of an experiment.

"It's not going to be that good," Rock warns the crowd, who are stunned at his unannounced arrival on the small comedy stage. "Not at these prices," he adds, joking "At these prices, I could leave right now!" He envisions the reviews: "Chris came out and he left. It was good! He didn't tell any jokes—but it was good!" Notes in hand, Rock warns the audience playfully that this isn't going to be a typical Chris Rock performance. Instead, he wants to work out new material under controlled conditions. "They'll give you about six minutes because you're famous," he explains. "... then you're back to square one." He wants to know what's funny, when he's not trying to be funny.

Rock's method is not unique. The Comedy Cellar is famous for big-name drop-ins: Dave Chappelle, Jon Stewart, and Amy Schumer are just a few comedians who have tested out their material in front of small crowds here before performing it on prime-time specials and in concert-scale gigs. Why perform at a small club when you can easily draw large crowds and thousands of dollars from a huge performance? Why show up unannounced and then deliberately undersell your own comedic abilities? What Rock and these other famous comedians recognize is the importance of the sixth principle of ultralearning: feedback.

THE POWER OF INFORMATION

Feedback is one of the most consistent aspects of the strategy ultralearners use. From the simple feedback of Roger Craig testing himself on *Jeopardy!* clues without knowing the answer to the uncomfortable feedback of Benny Lewis's approach of walking up to strangers to speak a language he only started learning the day prior, getting feedback was one of the most common tactics of the ultralearners I encountered. What often separated the ultralearning strategy from more conventional approaches was the immediacy, accuracy, and intensity of the feedback being provided. Tristan de Montebello could have taken the normal route of carefully preparing his script and then delivering a speech once every month or two, as is the case for most Toastmasters. Instead he dove straight in, speaking several times per week, jumping among different clubs to gather different perspectives on his performance. This deep dive into feedback was uncomfortable, but the rapid immersion also desensitized him to a lot of the anxiety that being onstage can create.

Feedback features prominently in the research on deliberate practice, a scientific theory of the acquisition of expertise initiated by K. Anders Ericsson and other psychologists. In his studies, Ericsson has found that the ability to gain immediate feedback on one's performance is an essential ingredient in reaching expert levels of performance. No feedback, and the result is often stagnation—long periods of time when you continue to use a skill but don't get any better at it. Sometimes the lack of feedback can even result in declining abilities. Many medical practitioners get worse with more experience as their accumulated knowledge from medical school begins to fade and the accuracy of their diagnoses is not given the rapid feedback that would normally promote further learning.

CAN FEEDBACK BACKFIRE?

The importance of feedback probably isn't too surprising; we all intuitively sense how getting information about what we're doing right and wrong can accelerate learning. More interestingly, the research on feedback shows that more isn't always better. Crucially, what matters is the type of feedback being given.

In a large meta-analysis, Avraham Kluger and Angelo DeNisi looked at hundreds of studies on the impact of providing feedback for learning. Though the overall effect of feedback was positive, it's important to note that in over 38 percent of cases, feedback actually had a negative impact. This leads to a confusing situation. On the one hand, feedback is essential for expert attainment, as demonstrated by the scientific studies of deliberate practice. Feedback also figures prominently in ultralearning projects, and it's difficult to imagine their being successful if their sources of feedback had been turned off. At the same time, a review of the evidence doesn't paint the picture of feedback being universally positive. What's the explanation?

Kluger and DeNisi argue that the discrepancy is in the type of feedback that is given. Feedback works well when it provides useful information that can guide future learning. If feedback tells you what you're doing wrong or how to fix it, it can be a potent tool. But feedback often backfires when it is aimed at a person's ego. Praise, a common type of feedback that teachers often use (and students enjoy), is usually harmful to further learning.

When feedback steers into evaluations of you as an individual (e.g., "You're so smart!" or "You're lazy"), it usually has a negative impact on learning. Further, even feedback that includes useful information needs to be correctly processed as a motivator and tool for learning. Kluger and DeNisi noted that some of the studies that showed a negative impact of feedback occurred because the subjects themselves chose not to use the feedback constructively. They may have rejected the feedback, lowered the standards they expect from themselves, or given up on the learning task altogether. The researchers note that who is giving the feedback can matter, as feedback coming from a peer or teacher has important social dynamics beyond mere information on how to improve one's abilities.

I find two things interesting about this research. First, it is clear that although informative feedback is beneficial, it can backfire if it is processed inappropriately or if it fails to provide useful information. This means that when seeking feedback, the ultralearner needs to be on guard for two possibilities. The first is overreacting to feedback (both positive and negative) that doesn't offer specific information that leads to improvement. Ultralearners need to be sensitive to what feedback is actually useful and tune out the rest. This is why, although all the ultralearners I met employed feedback, they didn't act on every piece of possible feedback. Eric Barone, for instance, did not attend to every comment and critique on early drafts of his game. In many cases he ignored them, when the feedback conflicted with his vision. Second, when it is incorrectly applied, feedback can have a negative impact on motivation. Not only can overly negative feedback lower your motivation, but so can overly positive feedback. Ultralearners must balance both concerns, pushing for the right level of feedback for their current stage of learning. Though we all know (and instinctively avoid) harsh and unhelpful criticism, the research also supports Rock's strategy of disregarding the positive feedback that his celebrity automatically generates.

The second interesting point about this research is that it explains why feedback-seeking efforts are often underused and thus remain a potent source of comparative advantage for ultralearners. Feedback is uncomfortable. It can be harsh and discouraging, and it doesn't always feel nice. Standing up on a stage in a comedy club to deliver jokes is probably one of the best ways to get better at stand-up comedy. But the act itself can be terrifying, as an awkward silence cuts deep. Similarly, speaking immediately in a new language can be painful, as the sense of your ability to communicate goes down precipitously from when you use your native tongue.

Fear of feedback often feels more uncomfortable than experiencing the feedback itself. As a result, it is not so much negative feedback on its own that can impede progress but the fear of hearing criticism that causes us to shut down. Sometimes the best action is just to dive straight into the hardest environment, since even if the feedback is very negative initially, it can reduce your fears of getting started on a project and allow you to adjust later if it proves too harsh to be helpful.

All of these acts require self-confidence, resolve, and persistence, which is why many self-directed learning efforts ignore seeking the aggressive feedback that could generate faster results. Instead of going to the source, taking feedback directly, and using that information to learn quickly, people often choose to dodge the punches and avoid a potentially huge source of learning. Ultralearners acquire skills quickly because they seek aggressive feedback when others opt for practice that includes weaker forms of feedback or no feedback at all.

WHAT KIND OF FEEDBACK DO YOU NEED?

Feedback shows up in many different forms for different types of learning projects. Getting good at stand-up comedy and learning to write computer programs involve very different kinds of feedback. Learning higher math and learning languages are going to use feedback in different ways. The opportunities for seeking better feedback will vary depending on what you're trying to learn. Rather than try to spell out exactly what feedback you need for your learning project, I think it's important to consider different types of feedback, along with how each one can be used and cultivated. By knowing what kind of feedback you're getting, you can make sure to use it best, while also recognizing its limitations. In particular, I want to consider three types of feedback: outcome feedback, informational feedback, and corrective feedback. Outcome feedback is the most common and in many situations the only type of feedback available. Informational feedback is also fairly common, and it's important to recognize when you can split apart outcomes to get feedback on parts of what you're learning and when feedback only on holistic outcomes is possible. Corrective feedback is the toughest to find but when employed well can accelerate learning the most.

Outcome Feedback: Are You Doing It Wrong?

The first type of feedback, and the least granular, is outcome feedback. This tells you something about how well you're doing overall but offers no ideas as to what you're doing better or worse. This kind of feedback can come in the form of a grade—pass/fail, A, B, or C—or it can come in the form of an aggregate feedback to many decisions you're making simultaneously.

The applause Tristan de Montebello received (or the crickets he heard) after a speech is an example of outcome feedback. It could tell him if he was getting better or worse, but it couldn't really say why or how to fix it. Every entrepreneur experiences this kind of feedback when a new product hits the market. It may sell wildly well or abysmally, but that feedback comes in bulk, not directly decomposable into the various aspects of the product. Did the product cost too much? Was the marketing message not clear enough? Was the packaging unappealing? Customer reviews and comments can provide clues, but ultimately the success or failure of any new product is a complex bundle of factors.

This type of feedback is often the easiest to get, and research shows that even getting this feedback, which lacks a specific message about what you need to improve, can be helpful. In one study, feedback for a task involving visual acuity facilitated learning, even when it was delivered in blocks that were too large to get any meaningful information about which responses were correct and which were incorrect. Many projects that wholly lack feedback can easily be changed to get this broad-scale feedback. Eric Barone, for instance, provided a development blog to publish work on his game and solicit feedback from early drafts. It couldn't provide him with detailed information about what exactly to improve and change, but his simply being immersed in an environment that provided feedback at all was helpful.

Outcome feedback can improve how you learn through a few different mechanisms. One is by providing you with a motivational benchmark against your goal. If your goal is to reach a certain quality of feedback, this feedback can give you updates on your progress. Another is that it can show you the relative merits of different methods you're trying. When you are progressing rapidly, you can stick to those learning methods and approaches. When progress stalls, you can see what you might be able to change in your current approach. Although outcome feedback isn't complete, it is often the only kind available and can still have a potent impact on your learning rate.

Informational Feedback: What Are You Doing Wrong?

The next type of feedback is informational feedback. This feedback tells you what you're doing wrong, but it doesn't necessarily tell you how to fix it. Speaking a foreign language with a native speaker who doesn't share a language with you is an exercise in informational feedback. That person's confused stare when you misuse a word won't tell you what the correct word is, but it will tell you that you're getting it wrong. Tristan de Montebello, in addition to the overall assessment of his performance by audience members at the end of a speech, can also get live informational feedback about how it's going moment to moment. Did that joke work? Is my story boring them? This is something you can spot in the distracted glances or background chatter throughout your speech. Rock's stand-up experiment is also a type of informational feedback. He can tell when a certain joke lands or doesn't, based on the reaction of the audience. However, they can't tell him what to do to make it funnier—he's the comedian, not them.

This kind of feedback is easy to obtain when you can get real-time access to a feedback source. A computer programmer who gets error messages when her programs don't compile properly may not have enough knowledge to understand what she's doing wrong. But as errors increase or diminish, depending on what she does, she can use that signal to fix her problems. Self-provided feedback is also ubiquitous, and in some pursuits it can be almost as good as feedback from others. When painting a picture, you can simply look at it and get a sense of whether your brushstrokes are adding to or detracting from the image you want to convey. Because this kind of feedback often comes from direct interaction with the environment, it often pairs well with the third principle, directness.

Corrective Feedback: How Can You Fix What You're Doing Wrong?

The best kind of feedback to get is corrective feedback. This is the feedback that shows you not only what you're doing wrong but how to fix it. This kind of feedback is often available only through a coach, mentor, or teacher. However, sometimes it can be provided automatically if you are using the right study materials. During the MIT Challenge, I did most of my practice by going back and forth between assignments and their solutions, so that when I finished a problem, I was shown not only whether I had gotten it right or wrong but exactly how my answer differed from the correct one. Similarly, flash cards and other forms of active recall provide corrective feedback by showing you the answer to a question after you make your guess.

The educators Maria Araceli Ruiz-Primo and Susan M. Brookhart argue, "The best feedback is *informative* and *usable* by the student(s) who receive it. Optimal feedback indicates the difference between the current state and the desired learning state AND helps students to take a step to improve their learning."

The main challenge of this kind of feedback is that it typically requires access to a teacher, expert, or mentor who can pinpoint your mistakes and correct them for you. However, sometimes the added edge of having corrective over merely informational feedback can be worth the effort needed to find such people. Tristan de Montebello worked with Michael Gendler to help him with his public speaking performance, and that helped him spot subtle weaknesses in his presentations that would have gone unnoticed by himself or by a less experienced audience member giving broader feedback.

This type of feedback trumps outcome feedback, which can't indicate what needs improving, and informational feedback, which can indicate what to improve but not how. However, it can also be unreliable. Tristan de Montebello would often get conflicting advice after delivering a speech; some audience members would tell him to slow down, while others said to speed up. This can also be a situation in which paying for a tutor can be useful, because that person can spot the exact nature of your mistake and correct it with less struggle on your part. The self-directed nature of ultralearning shouldn't convince you that learning is best done as an entirely solitary pursuit.

FURTHER NOTES ON TYPES OF FEEDBACK

A few things are worth noting here. First, you need to be careful when trying to "upgrade" feedback from a weaker form to a stronger form if it's not actually possible. To switch from outcome feedback to informational feedback, you need to be able to elicit feedback on a per element basis of what you're doing. If instead the feedback is being provided as a holistic assessment of everything you're doing, trying to turn it into informational feedback can backfire. Game designers know to watch out for this, because asking play testers what they don't like about a game can often return spurious results: for example, they don't like the color of the character or the background music. The truth is, the players are evaluating the game holistically, so they often can't offer this kind of feedback. If their responses come from using it as a whole, not from each aspect individually, asking for greater specificity may lead to guesses from those giving feedback.

Similarly, corrective feedback requires a "correct" answer or the response of a recognized expert. If there is no expert or a single correct approach, trying to turn informational feedback into corrective feedback can work against you when the wrong change is suggested as an improvement. De Montebello noted to me that the advice most people gave him wasn't terribly useful, but the consistency of it was. If his speech elicited wildly different reactions each time, he knew there was still a lot of work to do. When the speech started to get much more consistent comments, he knew he was onto something. This illustrates that ultralearning isn't simply about maximizing feedback but also knowing when to selectively ignore elements of it to extract the useful information. Understanding the merits of these different types of feedback, as well as the preconditions that make them possible, is a big part of choosing the right strategy for an ultralearning project.

HOW QUICK SHOULD FEEDBACK BE?

An interesting question in the research on feedback is how quick it should be. Should you get immediate information about your mistakes or wait some period of time? In general, research has pointed to immediate feedback being superior in settings outside of the laboratory. James A. Kulik and Chen-Lin C. Kulik review the literature on feedback timing and suggest that "Applied studies using actual classroom quizzes and real learning materials have usually found immediate feedback to be more effective than delay." Expertise researcher K. Anders Ericsson agrees, arguing in favor of immediate feedback when it assists in identifying and correcting mistakes and when it allows one to execute a corrected version of their performance revised in response to the feedback.

Interestingly, laboratory studies tend to show that delaying the presentation of the correct response along with the original task (delayed feedback) is more effective. The simplest explanation of this result is that presenting the question and answer again offers a second, spaced exposure to the information. If this explanation were correct, all it would mean is that that immediate feedback is best paired with delayed review (or further testing) to strengthen your memory compared with a single exposure. I'll cover more on spacing and how it impacts your memory in the next chapter on retention.

Despite the superficially mixed results on the timing of feedback from the scientific literature, I generally recommend faster feedback. This enables a quicker recognition of mistakes. However, there's a possible risk that this recommendation might backslide into getting feedback before you've tried your best to answer the question or solve the problem at hand. Early studies on feedback timing tended to show a neutral or negative impact of immediate feedback on learning. In those studies, however, experimenters often gave subjects the ability to see the correct answer before subjects had finished filling out the prompt. That meant subjects could often copy the correct answer rather than try to retrieve it. Feedback too soon may turn your retrieval practice effectively into passive review, which we already know is less effective for learning. For hard problems, I suggest setting yourself a timer to encourage you to think hard on difficult problems before giving up to look at the correct answer.

HOW TO IMPROVE YOUR FEEDBACK

By now you see the importance of feedback to your learning efforts. I've explained why feedback, especially when delivered to others, can sometimes backfire. I've also showed how the three types—outcome, informational, and corrective—have different strengths and the preconditions that need to be in place in order to make them effective. Now I want to focus on some concrete tactics you can apply to get better feedback.

Tactic 1: Noise Cancellation

Anytime you receive feedback, there are going to be both a signal—the useful information you want to process—and noise. Noise is caused by random factors, which you shouldn't overreact to when trying to improve. Say you're writing articles that you post online, trying to improve your writing ability. Most of them won't attract much attention, and when they do, it's often because of factors outside of your control; for example, just the right person happens to share it, causing it to spill across social networks. The quality of your writing does drive these factors, but there's enough randomness that you need to be careful not to change your entire approach based on one data point. Noise is a real problem when trying to improve your craft because you need to do far more work to get the same information about how to write well. By modifying and selecting the streams of feedback you pay attention to, you can reduce the noise and get more of the signal.

A noise-cancelling technique used in audio processing is filtering. Sound engineers know that human speech tends to fall within a particular range of frequencies, whereas white noise is all over the spectrum. They can boost the signal, therefore, by amplifying the frequencies that occur in human speech and quieting everything else. One way to do this is to look for proxy signals. These don't exactly equal success, but they tend to eliminate some of the noisy data. For blog writing, one way to do so would be to use tracking code to figure out what percentage of people read your articles all the way to the end. This doesn't prove your writing is good, but it's a lot less noisy than raw traffic data.

Tactic 2: Hitting the Difficulty Sweet Spot
Feedback is information. More information equals more opportunities to learn. A scientific measure of information is based on how easily you can predict what message it will contain. If you know that success is guaranteed, the feedback itself provides no information; you knew it would go well all along. Good feedback does the opposite. It is very hard to predict and thus gives more information each time you receive it.

The main way this impacts your learning is through the difficulty you're facing. Many people intuitively avoid constant failure, because the feedback it offers isn't always helpful. However, the opposite problem, of being too successful, is more pervasive. Ultralearners carefully adjust their environment so that they're not able to predict whether they'll succeed or fail. If they fail too often, they simplify the problem so they can start noticing when they're doing things right. If they fail too little, they'll make the task harder or their standards stricter so that they can distinguish the success of different approaches. Basically, you should try to avoid situations that always make you feel good (or bad) about your performance.

Tactic 3: Metafeedback

Typical feedback is performance assessment: your grade on a quiz tells you something about how well you know the material. However, there's another type of feedback that's perhaps even more useful: metafeedback. This kind of feedback isn't about your performance but about evaluating the overall success of the strategy you're using to learn.

One important type of metafeedback is your learning rate. This gives you information about how fast you're learning, or at least how fast you're improving in one aspect of your skill. Chess players might track their Elo ratings growth. LSAT studiers might track their improvements on mock exams. Language learners might track vocabulary learned or errors made when writing or speaking. There are two ways you can use this tool. One is to decide when you should focus on the strategy you're already using and when you should experiment with other methods. If your learning rate is slowing to a trickle, that might mean you're hitting diminishing returns with your current approach and could benefit from different kinds of drills, difficulties, or environments. A second way you can apply metafeedback is by comparing two different study methods to see which works better. During the MIT Challenge, I'd often split up questions from different subtopics before testing myself on an exam and try different approaches side by side. Does it work better to dive straight into trying to answer questions or to spend a little time to try to see that you understand the main concepts first? The only way you can know is to test your own learning rates.

Tactic 4: High-Intensity, Rapid Feedback

Sometimes the easiest way to improve feedback is simply to get a lot more of it a lot more often. This is particularly true when the default mode of learning involves little or infrequent feedback. De Montebello's strategy of improving public speaking relied largely on getting far more frequent exposure to the stage than most speakers do. Lewis's language immersion exposes him to information about his pronunciation at a point when most students still haven't uttered a word. High-intensity, rapid feedback offers informational advantages, but more often the advantage is emotional, too. Fear of receiving feedback can often hold you back more than anything. By throwing yourself into a high-intensity, rapid feedback situation, you may initially feel uncomfortable, but you'll get over that initial aversion much faster than if you wait months or years before getting feedback.

Being in such a situation also provokes you to engage in learning more aggressively than you might otherwise. Knowing that your work will be evaluated is an incredible motivator to do your best. This motivational angle for committing to high-intensity feedback may end up outweighing the informational advantage it provides.

BEYOND FEEDBACK

Receiving feedback isn't always easy. If you process it as a message about your ego rather than your skills, it's easy to let a punch become a knockout. Though carefully controlling the feedback environment so it is maximally encouraging may be a tantalizing option, real life rarely affords such an opportunity. Instead, it's better to get in and take the punches early so that they don't put you down for the count. Though short-term feedback can be stressful, once you get into the habit of receiving it, it becomes easier to process without overreacting emotionally.

Ultralearners use this to their advantage, exposing themselves to massive amounts of feedback so that the noise can be stripped away from the signal.

Feedback and the information it provides, however, is useful only if you remember the lessons it teaches. Forgetting is human nature, so it is not enough to learn; you also need to make the information stick. This brings us to the next principle of ultralearning, retention, in which we'll discuss strategies that will ensure the lessons you learn aren't forgotten.

CHAPTER X

PRINCIPLE 7

Retention

Don't Fill a Leaky Bucket

Memory is the residue of thought.
—*Daniel Willingham, cognitive psychologist*

In the small Belgian city of Louvain-la-Neuve, Nigel Richards has just won the World Scrabble Championships. On its own, this isn't too surprising. Richards has won a championship three times before, and both his prowess with the game and his mysterious personality have made him something of a legend in competitive Scrabble circles. This time, however, is different: instead of the original English-language version of the famous crossword game, Richards has won the French World Championship. This is a much harder feat: most English dictionary versions have roughly 200,000 valid word entries; French, with its gendered nouns and adjectives and copious conjugations, has nearly double that with around 386,000 valid word forms. To pull off such a feat is quite remarkable, even more so due to one simple fact: Richards doesn't speak French.

Richards, an engineer born and raised in Christchurch, New Zealand, is an unusual character. With his long beard and retro aviator sunglasses, he looks like a cross between Gandalf and Napoleon Dynamite. His skills at Scrabble, however, are no joke. A late starter to the game, his mother encouraged him to start in his late twenties, saying "Nigel, since you're no good at words, you won't be good at this game, but it will keep you occupied." From those inauspicious beginnings Richards has gone on to dominate the competitive Scrabble scene. Some people even argue that he may be the greatest player of all time.

In case you've been living under a rock, Scrabble is based on forming crosswords. Each player has seven letter tiles, drawn from a bag, with which to form words. The catch is that the words must link up with the words already on the board. To be a good player requires a voluminous memory, not only of the words we use every day but of obscure words that are useful because of their length or the letters they contain. A decent casual player quickly learns all the valid two-letter words, including unusual ones such as "AA" (a type of lava) and "OE" (a windstorm in the Faroe Islands). To perform at tournament level, however, requires memorizing nearly all of the short words, as well as longer seven- and eight-letter words, since if a player uses up all seven tiles in one turn, there is an extra fifty-point bonus (or "bingo," in Scrabble jargon). Memory, however, isn't the only skill needed. Like other competitive games, tournament Scrabble uses a timing system, so skilled players must not only be able to construct valid words from a scrambled set of tiles but quickly find spaces and calculate which words will score the most points. In this regard, Richards is a master: given the tiles CDHLRN and one blank (which can be used for any letter), Richards ignored the obvious CHILDREN and instead linked up multiple crosswords to make the even higher scoring CHLORODYNE.

Richards's virtuosity is only intensified by the mystery that surrounds it. He is quiet and mostly keeps to himself. He refuses all interviews with reporters and seems completely uninterested in fame, fortune, or even providing explanations for how he does it. A fellow competitor, Bob Felt, bumping into Richards at a tournament noted his monklike serenity, telling him "When I see you, I can never tell whether you've won or lost." "That's because I don't care" was Richard's monotone response. Even his competing in Belgium, which briefly pulled him into the international media spotlight, was done as an excuse to do a cycling trip through Europe. In fact, prior to his victory, he had spent only nine weeks preparing. After he beat a Francophone player, Schelick Ilagou Rekawe from Gabon, in the final match, he was given a standing ovation but needed a translator to thank the audience.

WHAT IS NIGEL RICHARDS'S SECRET?

The more I read about Nigel Richards, the more intrigued I became. Richards was as mysterious as he was incredible in his mnemonic abilities. He steadfastly ignores opportunities for interviews and is famously laconic in descriptions of his methods. After his victory in Louvain-la-Neuve, one reporter asked him if he had any special methods for memorizing all those words. "No" was Richards's monosyllabic response. Still, even if he wouldn't divulge his strategies publicly, perhaps some digging could reveal clues.

The first thing I discovered was that although Richards's victory in Belgium was astounding, it wasn't entirely without precedent. Other players of the game have won World Championships without being fluent in the language of competition. Scrabble is particularly popular in Thailand, for instance, and two former world champions, Panupol Sujjayakorn and Pakorn Nemitrmansuk, are not fluent in English. The reason is simple: remembering words in one's native language and remembering words in Scrabble are different mnemonic feats. In spoken language, the meaning of a word, its pronunciation, and its feel are important. In Scrabble, those things don't matter; words are just combinations of letters. Richards could win at French Scrabble without speaking French because the game wasn't much different from English; he just had to memorize different patterns of letters. A native speaker has an advantage, of course, since many spellings will already be familiar. But there will still be a large number of arcane and unfamiliar words to memorize, and the skill of rearranging the letters into valid board positions and calculating to achieve maximal points remains the same in every language in which Scrabble can be played.

The next piece of the puzzle I discovered was that Scrabble, it turns out, isn't the only activity in which Richards possesses a strange intensity. His other love is cycling. Indeed, in an early tournament in Dunedin, New Zealand, he got onto his bicycle after work finished, pedaled through the night from Christchurch to Dunedin, a distance of over two hundred miles, without sleeping, and started the tournament first thing in the morning. After he won, competitors he met at the tournament offered to give him a ride home. He politely declined, preferring to bicycle back the entire way home to Christchurch for another sleepless night before starting work again Monday morning. At first that felt like just another odd quirk in his profile, like his home-done haircuts and reluctance to be interviewed. Now, though, I believe it may hold some keys to unlocking some of his mystery.

Cycling, of course, isn't a great mnemonic technique. If it were, Lance Armstrong would have been a fierce contender. However, it does illustrate a common theme in Richards's personality that overlaps with that of other ultralearners I have encountered: an obsessive intensity that exceeds what is considered a normal investment of effort. Richards's cycling, it turns out, also lines up well with the only other clues I've been able to uncover about his methods: he reads lists; long lists of words, starting with two-letter words and then moving up. "The cycling helps," he explains, "I can go through lists in my mind." He reads the dictionary, focusing exclusively on combinations of letters, ignoring definitions, tenses, and plurals. Then, drawing from memory, he repeats them over and over again as he cycles for hours. This aspect also corresponds with a method that is common to other ultralearners and that has shown up in other principles of learning so far: active recall and rehearsal. By retrieving words, Richards likely takes his already impressive memory and makes it unassailable through active practice.

There are other clues about Richards's performance: he focuses on memory, not anagramming (rearranging the tiles to create words); he works forward and backward, starting from small words, going on to big ones and back again; he claims to recall the words visually, as he cannot remember words when they're spoken. All of these clues provide glimpses into Richards's mind, but they leave out even more than they reveal. How many times does he have to read the words from his list before he can rehearse it mentally? Are the words organized in some way or just listed alphabetically? Is he a savant with exceptional abilities and lower-than-normal general intelligence or an all-round genius for whom memorizing Scrabble words is just one of many impressive abilities? Maybe his intelligence is quite average and his dominance in Scrabble represents his extreme dedication to the game. We might never know the answers to those questions.

I certainly can't rule out the theory that Richards's mind is simply hardwired differently or better for memory than my own. After all, nothing I've encountered so far about his method is so boldly original that serious Scrabble players would be unaware of it. Yet Richards has completely dominated his competition. Part of me suspects that his intense, obsessive personality, which enables him to cycle for hours reviewing lists mentally, might also form at least a partial explanation. Whatever gifts he might possess, he also seems to possess the ultralearner ethos I've described thus far in the book. For whatever it is worth, Richards himself argues for more of the latter than the former: "It's hard work, you have to have dedication to learn," elsewhere adding "I'm not sure there is a secret, it's just a matter of learning the words."

Scrabble words may not be important to your life. However, memory is essential to learning things well. Programmers must remember the syntax for the commands in their code. Accountants need to memorize ratios, rules, and regulations. Lawyers must remember precedents and statutes. Doctors need to know tens of thousands of factoids, from anatomical descriptions to drug interactions. Memory is essential, even when it is wrapped up in bigger ideas such as understanding, intuition, or practical skill. Being able to understand how something works or how to perform a particular technique is useless if you cannot recall it. Retention depends on employing strategies so the things you learn don't leak out of your mind. Before discussing strategies of retention, however, let's take a look at why remembering things is so difficult.

WHY IS IT SO HARD TO REMEMBER THINGS?

Richards is an extreme case, but his story nonetheless illustrates many themes that are important for anyone who wants to learn something: How can you retain all of the things you learn? How can you defend against forgetting hard-won facts and skills? How can you store the knowledge you've acquired so that it can be easily retrieved exactly when you need it? In order to understand learning, you need to understand how and why you forget.

Losing access to previously learned knowledge has been a perennial problem for educators, students, and psychologists. Fading knowledge impacts the work you do as well. One study reported that doctors give worse medical care the longer they have worked, as their stored knowledge from medical school is gradually forgotten, despite their working in the profession full-time. Quoting from the original abstract:

> Physicians with more experience are generally believed to
> have accumulated knowledge and skills during years in
> practice and therefore to deliver high-quality care. However,
> evidence suggests that there is an inverse relationship
> between the number of years that a physician has been in
> practice and the quality of care that the physician provides.

Hermann Ebbinghaus, in one of the first psychological experiments in history, spent years memorizing nonsense syllables, much in the same way Richards memorizes Scrabble words, and carefully tracking his ability to recall them later. From this original research, later verified by more experimentally robust studies, Ebbinghaus discovered the forgetting curve. This curve shows that we tend to forget things incredibly quickly after learning them, there being an exponential decay in knowledge, which is steepest right after learning. However, Ebbinghaus noted, this forgetting tapers off, and the amount of knowledge forgotten lessens over time. Our minds are a leaky bucket; however, most of the holes are near the top, so the water that remains at the bottom leaks out more slowly.

Over the intervening years, psychologists have identified at least three dominant theories to help explain why our brains forget much of what we initially learn: decay, interference, and forgotten cues. Though the jury is still out on the exact mechanism underlying human long-term memory, these three ideas likely form some part of explaining why we tend to forget things and, conversely, provide insight into how we might better retain what we've learned.

Decay: Forgetting with Time

The first theory of forgetting is that memories simply decay with time. This idea does seem to match common sense. We remember events, news, and things learned in the past week much more clearly than things from last month. Things learned this year are recalled with much greater accuracy than events from a decade ago. By this understanding, forgetting is simply an inevitable erosion by time. Like sands in an hourglass, our memories inexorably slip away from us as we become more distant from them.

There are flaws with this theory being the complete explanation, however. Many of us can vividly recall events from early childhood, even if we can't remember what we ate for breakfast last Tuesday. There also seem to be patterns in which things are remembered and which are forgotten that go beyond the time since they were originally learned: vivid, meaningful things are more easily recalled than banal or arbitrary information. Even if there is a component to our forgetting that is simply decay, it seems exceedingly unlikely that this is the only factor.

Interference: Overwriting Old Memories with New Ones

Interference suggests a different idea: that our memories, unlike the files of a computer, overlap one another in how they are stored in the brain. In this way, memories that are similar but distinct can compete with one another. If you're learning programming, for instance, you may learn what a *for loop* is and remember it in terms of doing something repeatedly. Later, you may learn about *while loops, recursion, repeat-until loops*, and *go-to statements*. Now, each of these has to do with doing something repeatedly, but in different ways, so they may interfere with your ability to remember correctly what a *for loop* does. There are at least two flavors of this: proactive interference and retroactive interference. Proactive interference occurs when previously learned information makes acquiring new knowledge harder. Think of this as if the "space" where that information wants to be stored is already occupied, so forming the new memory becomes harder. This can happen when you want to learn the definition of a word but have difficulty because that word already has a different association in your mind. Consider trying to learn the concept of negative reinforcement in psychology—here the word "negative" has the meaning "absent," as opposed to "bad," so negative reinforcement is when you encourage a behavior by removing something, say a painful stimulus. However, since the earlier meaning of negative as "bad" already exists, you may have difficulty remembering this and it becomes easy to incorrectly equate negative reinforcement with punishment. Retroactive interference is the opposite—where learning something new "erases" or suppresses an old memory. Anyone who has learned Spanish and later tried to learn French knows how tricky retroactive interference can be, as French words pop out when you want to speak Spanish again.

Forgotten Cues: A Locked Box with No Key

The third theory of forgetting says that many memories we have aren't actually forgotten but simply inaccessible. The idea here is that in order to say that one has remembered something, it needs to be retrieved from memory. Since we aren't constantly experiencing the entirety of our long-term memories simultaneously, this means there must be some process for dredging up the information, given an appropriate cue. What may happen in this case is that one of the links in the chain of retrieving the information has been severed (perhaps by decay or interference) and therefore the entire memory has become inaccessible. However, if that cue were restored, or if an alternative path to the information could be found, we would remember much more than is currently accessible to us.

This explanation also has some advantages. Intuitively it seems to be somewhat true, as we all know the tip-of-the-tongue experience, when we feel as though we should be able to remember a fact or word but we're not able to summon it up immediately. It might also suggest that relearning things is much faster than learning them initially, because relearning is closer to repair work, while original learning is a completely new construction. Forgetting cues seems highly likely as a partial, if not complete, explanation of forgetting many things.

Cue forgetting as a complete explanation for our memory woes isn't without its problems, however. Many memory researchers now believe that the act of remembering is not a passive process. In recalling facts, events, or knowledge, we're engaging in a creative process of reconstruction. The memories themselves are often modified, enhanced, or manipulated in the process of remembering. It may be, then, that "lost" memories that are retrieved through new cues are actually fabrications. This seems especially likely in the case of "recovered" witness testimony from traumatic events, as experiments have shown that even highly vivid memories that feel completely authentic to the subject can be untrue.

HOW CAN YOU PREVENT FORGETTING?

Forgetting is the default, not the exception, so the ultralearners I encountered had devised various strategies for coping with this fact of life. These methods roughly divide into tackling two similar but different problems. The first set of methods deals with the problem of retention while undertaking the ultralearning project: How can you retain the things you learned the first week, so that you don't need to relearn them by the last week? This is particularly important for memory-intensive ultralearning efforts such as Benny Lewis's language learning and Roger Craig's *Jeopardy!* trivia mastery. In these domains and many others, the volume of information to be learned is often so large that the forgetting becomes a practical obstacle almost immediately. The second set of methods, in contrast, has to do with the longevity of the skills and knowledge acquired after the project has been completed: Once a language has been learned to a level you're satisfied with, how can you keep yourself from forgetting it completely a couple years later?

The ultralearners I encountered had devised differing methods for dealing with these two problems, which varied in effort and intensity. Some, like Craig, preferred elaborate electronic systems that can optimize memory with fancy algorithms, leaving little waste and inefficiency, if at the cost of introducing greater complexity. Others, like Richards, seem to prefer basic systems that succeed on their simplicity.

You need to pick a mnemonic system, which will both accomplish your goals and be simple enough to stick to. During intense periods of language learning, the sheer volume of vocabulary often meant that spaced-repetition systems were helpful for me. Other times, I preferred having conversations to maintain my speaking ability, even though this method is not quite as precise. With other subjects, I'm happier to allow for some degree of forgetting as long as I practice the skills I need to use continuously and have the ability to relearn.

My approaches may not reach a theoretical ideal, but they may end up working better because they have fewer possibilities for error and can be sustained more easily. Regardless of the exact system used, however, all systems seemed to work according to one of four mechanisms: spacing, proceduralization, overlearning, or mnemonics. Let's look at each of these mechanisms of retention first, in order to make sense of the quite different and idiosyncratic manifestations used in different ultralearning projects.

Memory Mechanism 1—Spacing: Repeat to Remember
One of the pieces of studying advice that is best supported by research is that if you care about long-term retention, don't cram.

Spreading learning sessions over more intervals over longer periods of time tends to cause somewhat lower performance in the short run (because there is a chance for forgetting between intervals) but much better performance in the long run. This was something I needed to be careful about during the MIT Challenge. After my first few classes, I switched from doing one class at a time to doing a few in parallel, to minimize the impact that the crammed study time would have on my memory.

If you have ten hours to learn something, therefore, it makes more sense to spend ten days studying one hour each than to spend ten hours studying in one burst. Obviously, however, if the amount of time between study intervals gets longer and longer, the short-term effects start to outweigh the long-term ones. If you learn something with a decade separating study intervals, it's quite possible that you'll completely forget whatever you had learned before you reach the second session.

Finding the exact trade-off point between too long and too short has been a minor obsession for some ultralearners. Space your study sessions too closely, and you lose efficiency; space them too far apart, and you forget what you've already learned. This has led many ultralearners to apply what are known as spaced-repetition systems (SRS) as a tool for trying to retain the most knowledge with the least effort. SRS was a major force behind Roger Craig's *Jeopardy!* trivia memorization, and I used the systems extensively when learning Chinese and Korean. Although you may not have heard of this term, the general principle is the backbone of many language-learning products, including Pimsleur, Memrise, and Duolingo. These programs tend to hide the spacing algorithm in the background, so you don't need to bother yourself with it. However, other programs, such as the open-source Anki, are the preferred tool of more extreme ultralearners who want to squeeze out a little more performance.

SRS is an amazing tool, but it tends to have quite focused applications. Learning facts, trivia, vocabulary words, or definitions is ideally suited for flash card software, which presents knowledge in terms of a question with a single answer. It's more difficult to apply to more complicated domains of knowledge, which rely on complex information associations that are built up only through real-world practice. Still, for some tasks, the bottleneck of memory is so tight that SRS is a powerful tool for widening it, even if there are some drawbacks. The authors of a popular study guide for medical students center their approach around SRS, because a medical student must remember so many things and the default strategy of forgetting and relearning is quite costly in terms of time.

Spacing does not require complex software, however. As Richards's story clearly demonstrates, simply printing lists of words, reading them over, and then rehearsing them mentally without having them in front of you is an incredibly powerful technique. Similarly, semiregular practice of a skill is often quite helpful. After my year of learning languages, I wanted to ensure that I didn't forget them. My approach was fairly simple: schedule thirty minutes of conversation practice once a week, to be done over Skype using italki, an online service for tutoring and language exchange partners all over the world. I maintained this for one year, after which I dropped to once-per-month practice for another two years. I don't know whether this practice schedule was ideal, and I had other opportunities to practice that came up spontaneously in that time period that also helped, but I believe it was much better than doing nothing and letting the skills atrophy. When it comes to retention, don't let perfect become the enemy of good enough.

Another strategy for applying spacing, which can work better for more elaborate skills that are harder to integrate into your daily habits, is to semiregularly do refresher projects. I leaned toward this approach for the things I learned during the MIT Challenge, since the skill I wanted most to retain was writing code, which is tricky to do on only half an hour per week. This approach has the disadvantage of sometimes deviating quite a lot from optimal spacing; however, if you're prepared to do a little bit of relearning to compensate, it can still be a better approach than completely giving up practice. Scheduling this kind of maintenance in advance can also be helpful, as it will remind you that learning isn't something done once and then ignored but a process that continues for your entire life.

Memory Mechanism 2—Proceduralization: Automatic Will Endure

Why do people say it's "like riding a bicycle" and not "like remembering trigonometry?" This common expression may be rooted in deeper neurological realities than it first appears. There's evidence that procedural skills, such as riding a bicycle, are stored in a different way from declarative knowledge, such as knowing the Pythagorean Theorem or the Sine Rule for triangles. This difference between *knowing how* and *knowing that* may also have different implications for long-term memory. Procedural skills, such as the ever-remembered bicycling, are much less susceptible to being forgotten than knowledge that requires explicit recall to retrieve.

This finding can actually be used to our advantage. One

dominant theory of learning suggests that most skills proceed through stages—starting declarative but ending up procedural as you practice more. A perfect example of this declarative-to-procedural transition is typewriting. When you start typing on a keyboard, you must memorize the positions of the letters. Each time you want to type a word, you have to think in terms of its letters, recall each one's position on the keyboard, and then move your finger to that spot to press it. This process may fail; you may forget where a key is and need to look down to type it. However, if you practice more and more, you stop having to look down. Eventually you stop having to think about the letters' positions or how to move your fingers to meet them. You may even reach a point where you don't think of letters at all and whole words come out at a time. Such procedural knowledge is quite robust and tends to be retained much longer than declarative knowledge. A quick observation is enough to verify this: when you've gotten really good at typing and someone asks you to quickly say where on a keyboard the letter *w* is, you might need to actually put your hands in the keyboard position (or imagine you're doing so) and pretend to type the *w* to say definitively. This is exactly what happened to me as I was typing out this paragraph. What has happened is that what was originally the primary access point to knowledge, your explicit memory of the key location, has faded away and now needs to be recalled with the more durable procedural knowledge encoded in your motor movements. If you've ever had to enter a password or pin code you use often, you may be in a similar position, where you remember it by feel and not by its explicit combination of numbers and letters.

Because of the fact that procedural knowledge is stored for longer, this may suggest a useful heuristic. Instead of learning a large volume of knowledge or skills evenly, you may emphasize a core set of information much more frequently, so that it becomes procedural and is stored far longer. This was an unintentional side effect of my friend's and my language-learning project. Being forced to speak a language constantly meant that a core set of phrases and patterns was repeated so often that neither of us will ever forget them. This may not hold true for a bunch of less frequently used words or phrases, but the starting points of conversations are nearly impossible to forget. The classic approach to language studies, in which students "move on" from beginner words and grammatical patterns to more complicated ones may sidestep this, so that those core patterns aren't sticky enough to last for years without repeated practice.

Failing to fully proceduralize core skills was a major flaw of my first major self-education effort, the MIT Challenge, which I was able to improve upon in my subsequent language-learning and portrait-drawing projects. Whereas the MIT Challenge did have core mathematical and programming skills that were often repeated, what ended up being proceduralized was more haphazard rather than reflecting a conscious decision to automate the most essential skills of applying computer science.

Most skills we learn are incompletely proceduralized. We may be able to do some of them automatically, but other parts require us to actively search our minds. You might, for instance, be able to easily move variables from one side of an equation to the other in algebra without thinking. But you may have to think a bit more when exponents or trigonometry is involved. Perhaps, owing to their nature, some skills cannot be completely automated and will always require some conscious thought. This creates an interesting mix of knowledge, with some things retained quite stably over longer periods of time and others susceptible to being forgotten. One strategy for applying this concept might be to ensure that a certain amount of knowledge is completely pro-ceduralized before practice concludes. Another approach might be to spend extra effort to proceduralize some skills, which will serve as cues or access points for other knowledge. You may aim to completely proceduralize the process you use to start work-ing on a new programming project, for example, so that you can get over that hump in the process of writing a new program. These strategies are somewhat speculative, but I think there are lots of potential ways the declarative-to-procedural transition of knowledge might be applied by clever ultralearners in the future.

Memory Mechanism 3—Overlearning: Practice Beyond Perfect

Overlearning is a well-studied psychological phenomenon that's fairly easy to understand: additional practice, beyond what is re-quired to perform adequately, can increase the length of time that memories are stored. The typical experimental setup is to give subjects a task, such as assembling a rifle or going through an emergency checklist, allowing them enough time to practice that they can do it correctly once. The time from zero to this point is considered the "learning" phase. Next, allow the subjects dif-ferent amounts of "overlearning," or practice that continues af-ter the first correct application. Since subjects are already doing the skill correctly, performance doesn't improve past this point. However, the overlearning can extend the durability of the skill.

In the typical setting in which overlearning has been studied, the duration of the overlearning effects tends to be quite short; practicing a little longer in one session produces an additional week or two of recall. This may imply that overlearning is primarily a short-term phenomenon: something useful for skills like first aid or emergency response protocols, which are rarely practiced but need to be kept fresh in between regular training sessions. I suspect, however, that overlearning might have longer-term implications if it is combined with spacing and proceduralization over much longer projects. In my own personal experience drawing portraits, for instance, the thought process used for mapping out the facial features I learned from Vitruvian Studio was repeated so many times that it's hard to forget, even though my major practice time was only during one month. Similarly, certain reflexes of programming or mathematics I can still easily recall from my MIT Challenge days, even without practice in the interim, because they happened to be patterns that were repeated far more than was necessary to perform them adequately at the time (because they were components of more elaborate problems).

Overlearning dovetails nicely with the principle of directness. Because direct use of a skill frequently involves overpracticing certain core abilities, that kernel is usually quite resistant to forgetting, even years later. In contrast, academically learned subjects tend to distribute practice more evenly to cover the entire curriculum to a minimum level of competency in each area, regardless of the centrality of subtopics to practical applications. Many people I've known who have learned a language that I also speak but who learned it through years of formal schooling have much more impressive vocabularies or knowledge of grammatical nuances than I do. However, those same people may trip over fairly basic phrases, because they learned every fact and skill evenly, rather than overlearning the smaller subset of very common patterns.

There seem to be two main methods I've encountered for applying overlearning. The first is core practice, continually practicing and refining the core elements of a skill. This approach often works well paired with some kind of immersion or working on extensive (as opposed to intensive) projects, after the initial ultralearning phase has been completed. The shift from learning to doing here may actually involve a deeper, subtler form of learning, which shouldn't be discounted as simply applying previously learned knowledge.

The second strategy is advanced practice, going one level above a certain set of skills so that core parts of the lower-level skills are overlearned as one applies them in a more difficult domain. One study of algebra students demonstrated this second strategy. Most students who had taken an algebra class and were retested years later had forgotten huge amounts of what they had learned. This could have been either because the information was truly lost or simply because forgotten cues rendered the majority of it inaccessible. Interestingly, this rate of forgetting was the same for better- and poorer-performing students; better students retained more than weaker ones, but the rate at which they had forgotten was the same. One group, however, did not show such a steep decline in forgetting: those who had taken calculus. This suggests that moving up a level to a more advanced skill enabled the earlier skill to be overlearned, thus preventing some forgetting.

Memory Mechanism 4—Mnemonics:
A Picture Retains a Thousand Words
The final tool common to many ultralearners I encountered was mnemonics. There are many mnemonic strategies, and covering them all is outside the scope of this book. What they have

in common is that they tend to be hyperspecific—that is, they are designed to remember very specific patterns of information. Second, they usually involve translating abstract or arbitrary information into vivid pictures or spatial maps. When mnemonics work, the results can be almost difficult to believe. Rajveer Meena, the Guinness World Record holder for memorizing digits of the mathematical constant *pi*, knows the number to 70,000 decimal places. Master mnemonicists, who compete in championships of memory, can memorize the order of a deck of cards in under sixty seconds and can repeat a poem verbatim after only a minute or two of studying. These feats are quite impressive, and even better, they can be learned by anyone patient enough to apply them. How do they work?

One common, and useful, mnemonic is known as the keyword method. The method works by first taking a foreign-language word and converting it into something it sounds like in your native language. If I were doing this with French, for example, I might take the word *chavirer* (to capsize) and convert it into "shave an ear," to which it is close enough in sound for the latter to serve as an effective cue for recalling the original word. Next I create a mental image that combines the sounds-like version of the foreign word and an image of its translation in a fantastical and vivid setting that is bizarre and hard to forget. In this case, I could imagine a giant ear shaving a long beard while sitting in a boat that capsizes. Then, whenever I need to remember what "capsize" is in French, I think of capsizing, recall my elaborate picture, which links to "shaving an ear" and thus ... *chavirer.* This process sounds needlessly complicated and elaborate at first, but it benefits from converting a difficult association (between arbitrary sounds and a new meaning) into a few links that are much easier to associate and remember. With practice, each conversion of this type may take only fifteen to twenty seconds, and it really does help with remembering foreign-language words. This particular kind of mnemonic works for this purpose, but there are others that work for remembering lists, numbers, maps, or sequences of steps in a procedure. For a good introduction to this topic, I highly recommend Joshua Foer's book *Moonwalking with Einstein: The Art and Science of Remembering Everything.*

Mnemonics work well, and with practice, anyone can do them. Why, then, are they not front and center in this chapter, instead of at the end? I believe that mnemonics, like SRS, are incredibly powerful tools. And as tools, they can open new possibilities for people who are not familiar with them. However, as someone who has spent much time exploring them and applying them to real-world learning, their applications are quite a bit narrower than they first appear, and in many real-world settings they simply aren't worth the hassle.

I believe there are two disadvantages to mnemonics. The first is that the most impressive mnemonics systems (like the one for memorizing thousands of digits of the mathematical constant *pi*), also require a considerable up-front investment. After you're done, you can memorize digits easily, but this isn't actually a very useful task. Most of our society adapts around the fact that people generally cannot memorize digits, so we have paper and computers do it for us. The second disadvantage is that recalling from mnemonics is often not as automatic as directly remembering something. Knowing a mnemonic for a foreign-language word is better than failing to remember it entirely, but it's still too slow to allow you to fluently form sentences out of mnemonically remembered words. Thus mnemonics can act as a bridge for difficult-to-remember information, but it's usually not the final step in creating memories that will endure forever.

Mnemonics, therefore, are an incredibly powerful if somewhat brittle tool. If you are doing a task that requires memorizing highly dense information in a very specific format, especially if the information is going to be used over a few weeks or months, they can enable you to do things with your mind that you might not have thought possible. Alternatively, they can be used as an intermediate strategy to smooth initial information acquisition when the information is quite dense. I've found them useful for language learning and terminology, and, paired with SRS, they can form an effective bridge from feeling as though there's no way you can possibly remember everything to remembering it so deeply that you can't possibly forget. Indeed, in a world before paper, computers, and other externalized memories, mnemonics were the main game in town. However, in the modern world, which has developed excellent coping mechanisms for the fact that most people cannot remember things as a computer can, I feel that mnemonics tend to serve more as cool tricks than as a foundation you should base your learning efforts on. Still, there is a devoted subset of ultralearners who are fiercely committed to applying these techniques, so my word shouldn't be the final verdict.

WINNING THE WAR AGAINST FORGETTING

To retain knowledge is ultimately to combat the inevitable human tendency to forget. This process occurs in all of us, and there's no way to avoid it completely. However, certain strategies—spacing, proceduralization, overlearning, and mnemonics—can counteract your short- and long-term rates of forgetting and end up making a huge difference in your memorization.

I opened this chapter by discussing Nigel Richards's mysterious Scrabble mastery. How he is able to recall so many words so quickly and see them in a set of scrambled tiles will likely remain an enigma. What we do know about him fits the picture of other ultralearners who have dominated memory-intensive subjects: active recall, spaced rehearsal, and an obsessive commitment to intense practice. Whether you or I have the will to go as far as Richards does is an open question, but with hard work and a good strategy, it seems likely to me that the battle against forgetting need not be a losing one.

Though Richards's Scrabble practice may give him the benefit of memorizing words he doesn't know the meaning of, real life tends to reward a different kind of memory: one that integrates knowledge into a deep understanding of things. In the next principle, we'll look at going from memory to intuition.

CHAPTER XI

PRINCIPLE 8

Intuition

Dig Deep Before Building Up

> Do not ask whether a statement is true until
> you know what it means.
> —*Errett Bishop, mathematician*

To the world, he was an eccentric professor and Nobel Prize–winning physicist; to his biographer, he was a genius; but to those who knew him, Richard Feynman was a magician. His colleague the mathematician Mark Kac once posited that the world holds two types of geniuses. The first are ordinary geniuses: "Once we understand what they have done we feel certain that we, too, could have done it." The other type are magicians, whose minds work in such inscrutable ways that "Even after we understand what they have done, the process by which they have done it is completely dark." Feynman, by his reckoning, was "a magician of the highest caliber."

Feynman could take problems others had worked on for months

and immediately see the solution. In high school, he competed in mathematics tournaments, where he would often get the correct answer while the problem was still being stated. While his competitors had just begun to compute, Feynman already had the answer circled on the page. In his college days, he competed in the Putnam Mathematics Competition, with the winner receiving a paid scholarship to Harvard. This competition is notoriously difficult, requiring clever tricks rather than straightforward application of previously learned principles. Time is also a factor, and some examination sessions have a median score of zero, meaning the typical competitor didn't get even one right. Feynman walked out of the exam early. He scored first place, with his fraternity brothers later being amazed at the drastic gap between Feynman's score and the next four on the list. During his work on the Manhattan Project, Niels Bohr, then one of the most famous and important living physicists, asked to speak with Feynman directly, to run his ideas by the young grad student before talking with the other physicists. "He's the only guy who's not afraid of me" was Bohr's explanation. "[He] will say when I've got a crazy idea."

Nor was Feynman's magic restricted to physics. As a child he went around fixing people's radios, in part because paying an adult for repairs in the Depression was too costly but also because the radio owners marveled at his process. Once, while he was lost in thought trying to figure out why a radio was producing an awful noise as it started up, the owner of the radio got impatient. "What are you doing? You come to fix the radio, but you're only walking back and forth!" "I'm thinking!" came the reply, at which the owner, startled at the boldness for which Feynman would later become famous, laughed. "He fixes radios by thinking!"

As a young man during the construction of the atomic bomb in the Manhattan Project, he occupied his free time picking the locks of his supervisors' desks and cabinets. He once broke into a senior colleague's filing cabinet, where the secrets for building a nuclear bomb were kept, as a practical joke. Another time, he demonstrated his technique to a military official, who, instead of fixing the security flaw, decided the proper course was to warn everyone to keep Feynman away from their safes! Later, upon meeting a locksmith, he found that his reputation had grown to the point where the professional said, "God! You're Feynman—the great safecracker!"

He also created the impression of being a human calculator. On a trip to Brazil, he went toe to toe against an abacus salesman, computing difficult figures such as the cube root of 1,729.03. Not only did Feynman get the right answer, 12.002, but he got it to more decimal places than the abacus salesman, who was still furiously calculating to get to 12 when Feynman displayed his five-digit result. This ability impressed even other professional mathematicians, to whom he argued that he could, within one minute, get the answer to any problem that could be stated in ten seconds to within 10 percent of the correct number. The mathematicians threw questions at him such as "e to the power of 3.3" or "e to the power of 1.4," and Feynman managed to spit back the correct answer almost immediately.

DEMYSTIFYING FEYNMAN'S MAGIC

Feynman was certainly a genius. Many people, including his biographer James Gleick, are satisfied to leave it at that. A magic trick, after all, is most dazzling when you don't know how it is done. Perhaps this is why many accounts of the man have focused on his magic instead of his method.

Though Feynman was quite smart, his magic had its gaps. He

excelled in math and physics but was abysmal in the humanities. His college grades in history were in the bottom fifth of his class, in literature in the bottom sixth, and his fine arts grades were worse than those of 93 percent of his fellow students. At one point, he even resorted to cheating on a test to pass. His intelligence, measured while he was in school, scored 125. The average college graduate has a score of 115, which puts Feynman only modestly higher. Perhaps, as has been argued afterward, Feynman's genius failed to be captured in his IQ score, or it simply was a poorly administered test. However, for someone so celebrated for a mind beyond comprehension, these facts remind us that Feynman was mortal.

What about Feynman's mental calculus? In this case, we have Feynman's words himself for how he could compute so much faster than the abacus salesman or his mathematician colleagues. The cube root of 1,729.03? Feynman explained, "I happened to know that a cubic foot contains 1728 cubic inches, so the answer is a tiny bit more than 12. The excess, 1.03, is only one part in nearly 2000, and I had learned in calculus that for small fractions, the cube root's excess is one-third of the number's excess. So all I had to do was find the fraction 1/1728, and multiply by 4." The constant e to the power of 1.4? Feynman revealed, "because of radioactivity (mean-life and half-life), I knew the log of 2 to the base e, which is .69315 (so I also knew that e to the power of .7 is nearly equal to 2)." To go to the power of 1.4, he'd just have to multiply that number against itself. "[S]heer luck," he explained. The secret was his impressive memory for certain arithmetic results and an intuition with numbers that enabled him to interpolate. However, the lucky picks of his examiners allowed him to leave an impression of a magical ability to calculate.

How about the famous lock picking? Once again, it was magic, in the same sense as a magician performing well-practiced tricks.

He obsessed over figuring out how combination locks worked. One day he realized that by fiddling with a lock when it was open, he could figure out the last two numbers on the safe. He would write them down on a note after he left the person's office and then could sneak back in, crack the remaining number with some patience, and leave ominous notes behind.

Even his magical intuition for physics had its explanation: "I had a scheme, which I still use today when somebody is explaining something that I'm trying to understand: I keep making up examples." Instead of trying to follow an equation, he would try to imagine the situation it described. As more information was given, he'd work it through on his example. Then whenever his interlocutor made a mistake, he could see it. "As they're telling me the conditions of the theorem, I construct something which fits all the conditions. You know, you have a set (one ball)— disjoint (two balls). Then the balls turn colors, grow hairs, or whatever, in my head as they put more conditions on. Finally they state the theorem, which is some dumb thing about the ball which isn't true for my hairy green ball thing, so I say, 'False!'"

Magic, perhaps, Feynman did not possess, but an incredible intuition for numbers and physics he certainly did. This might downplay the idea that his mind worked in a fundamentally different way from yours or mine, but it doesn't negate the impressiveness of his feats. After all, even knowing the logic behind Feynman's sleight of hand, I'm certain I wouldn't have been able to calculate the numbers he did so effortlessly or follow some complex theory in my mind's eye. This explanation doesn't provide the satisfying "Aha!" that it would have had the magician's trick been revealed as something trivial. Therefore, we need to dig deeper to an understanding of how someone such as Feynman could develop this incredible intuition in the first place.

INSIDE THE MIND OF THE MAGICIAN

Psychological researchers have investigated the problem of how intuitive experts, such as Feynman, think differently about problems than novices do. In a famous study, advanced PhDs and undergraduate physics students were given sets of physics problems and asked to sort them into categories. Immediately, a stark difference became apparent. Whereas beginners tended to look at superficial features of the problem—such as whether the problem was about pulleys or inclined planes—experts focused on the deeper principles at work. "Ah, so it's a conservation of energy problem," you can almost hear them saying as they categorized the problem by what principles of physics they represented. This approach is more successful in solving problems because it gets to the core of how the problems work. The surface features of a problem don't always relate to the correct procedure needed to solve it. The students needed much more trial and error to home in on the correct method, whereas the experts could immediately start with the right approach.

If the principles-first way of thinking of problems is so much more effective, why don't students start there instead of attending to superficial characteristics? The simple answer may be that they can't. Only by developing enough experience with problem solving can you build up a deep mental model of how other problems work. Intuition sounds magical, but the reality may be more banal—the product of a large volume of organized experience dealing with the problem.

Another study, this time comparing chess masters and beginners, offered an explanation of why this might be so. The memory for chess positions of experts and novices was tested by showing them a particular chess setup and then asking them to re-create it on an empty board. The masters could recall far more than the beginners. The new players needed to put down pieces one by one and were often unable to fully remember all the details of the position. The masters, in contrast, remembered the board in larger "chunks" with several pieces corresponding to a recognizable pattern put down at the same time. Psychologists theorize that the difference between grand masters and novices is not that grand masters can compute many more moves ahead but that they have built up huge libraries of mental representations that come from playing actual games. Researchers have estimated that having around 50,000 of these mental "chunks" stored in long-term memory is necessary to reach expert status. These representations allow them to take a complex chess setup and reduce it to a few key patterns that can be worked with intuitively. Beginners, who lack this ability, have to resort to representing each piece as a single unit and are therefore much slower.*

This facility of chess grand masters, however, is limited to the patterns that come from real chess games. Give beginners and experts a randomized chess board (one that doesn't arise from normal play), and the experts no longer display the same marked advantage. Without the library of memorized patterns at their disposal, they have to resort to the beginner's method of remembering the board piece by piece.

This research gives us a glimpse into how the mind of a great intuitionist such as Feynman operated. He, too, focused on principles first, building off examples that cut straight to the heart of

---

* It should be noted that not all researchers agree with the chunking model. K. Anders Ericsson, the psychologist behind deliberate practice, prefers an alternative model called "Long-Term Working Memory." The differences are largely technical, and both models point to the idea of expertise being gained through extensive context-specific practice.

what the problem represented rather than focusing on superficial features. His ability to do this was also built off an impressive library of stored physics and math patterns. His mental calculation feats seem impressive to us but were trivial to him, because he happened to know so many mathematical patterns. Like chess grand masters, when given real physics problems he excelled because he had built a huge library of patterns from real experiences with physics. However, his intuition, too, would fail him when the subject of his study wasn't built on those assumptions. Feynman's mathematician friends would test him on counterintuitive theorems from mathematics. His intuition there would fail when properties of the procedure (such as that an object can be cut into infinitely small pieces) defied the normal physical limitations that aided his intuition elsewhere.

Feynman's magic was his incredible intuition, coming from years of playing with the patterns of math and physics. Could emulating his approach to learning enable someone else to capture some of that magic? Let's look at some of Feynman's hallmark approaches to learning and solving problems and try to reveal some of the magician's secrets.

HOW TO BUILD YOUR INTUITION

Simply spending a lot of time studying something isn't enough to create a deep intuition. Feynman's own experience demonstrates this. On numerous occasions, he would encounter students who memorized solutions to a particular problem but failed to see how they applied outside the textbook domain. In one story, he tricked some of his classmates into believing that a French curve (a device for drawing curved lines) was special because, no matter how you hold it, the bottom is tangent to a horizontal line. This, however, is true of any smooth shape, and it is an elementary fact of calculus that his fellow classmates should have realized. Feynman saw this as an example of a particularly "brittle" way of learning things, since students didn't really think about relating what they had learned to problems outside the textbook.

How, then, can someone avoid a similar fate—spending a lot of time learning something without really developing the flexible intuition for it that made Feynman famous? There's no precise recipe for doing so, and a healthy dose of experience and smarts certainly helps. However, Feynman's own account of his learning process offers some useful guidelines for how he did things differently.

Rule 1: Don't Give Up on Hard Problems Easily

Feynman was obsessed with solving problems. Starting in his childhood days of tinkering with radios, he would work stubbornly on a problem until it yielded. Sometimes, when the owner of the radio would get impatient, he recalled, "If [he] had said, 'Never mind, it's too much work,' I'd have blown my top, because I want to beat this damn thing, as long as I've gone this far." This tendency carried over into mathematics and physics. He'd often eschew easier methods, such as the Lagrangian technique, forcing himself to painstakingly calculate all the forces by hand, simply because with the latter method he came to understand it better. Feynman was a master at pushing farther on problems than others expected of him, and this itself might have been the source of many of his unorthodox ideas.

One way you can introduce this into your own efforts is to give yourself a "struggle timer" as you work on problems. When you feel like giving up and that you can't possibly figure out the solution to a difficult problem, try setting a timer for another ten minutes to push yourself a bit further. The first advantage of this struggle period is that very often you can solve the problem you are faced with if you simply apply enough thinking to it. The second advantage is that even if you fail, you'll be much more likely to remember the way to arrive at the solution when you encounter it. As mentioned on the chapter on retrieval, difficulty in retrieving the correct information—even when the difficulty is caused by the information not being there—can prime you to remember information better later.

Rule 2: Prove Things to Understand Them

Feynman told a story of his first encounter with the work by the physicists T. D. Lee and C. N. Yang. "I can't understand these things that Lee and Yang are saying. It's all so complicated," he declared. His sister, lightly teasing him, remarked that the problem wasn't that he couldn't understand it but that he hadn't invented it. Afterward, Feynman decided to read through the papers meticulously, finding that they weren't actually so difficult but that he had simply been afraid to go through them.

Though this story illustrates one of Feynman's quirks, it is also revealing because it illustrates a major point in his method. Feynman didn't master things by following along with other people's results. Instead, it was by the process of mentally trying to re-create those results that he became so good at physics. This could be a disadvantage at times, since it caused him to repeat work and reinvent processes that already existed in other forms. However, his drive to understand things by virtue of working through the results himself also assisted in building his capacity for deep intuition.

Feynman was not alone in this approach. Albert Einstein, as a child, built his intuitive powers by trying to prove propositions in math and physics. One of Einstein's earliest mathematical forays was trying to prove the Pythagorean Theorem on the basis of similar triangles. What this approach indicates is that both men had a tendency to dig much deeper before they considered something to be "understood." Feynman's scoffing at not understanding Lee and Yang wasn't because there was no understanding; indeed, he was familiar with much of the background work on the problem. Instead, it was probably because his notion of understanding was much deeper and more based on demonstrating results himself, rather than merely nodding along while reading.

The challenge of thinking you understand something you don't is unfortunately a common one. Researcher Rebecca Lawson calls this the "illusion of explanatory depth." At issue here is the notion that we judge our own learning competency, not directly but through various signals. Assessing whether or not we know a factual matter, such as what is the capital of France, is quite easy—either the word "Paris" comes up in your mind, or it doesn't. Asking whether you understand a concept is a lot harder because you may understand it a little, but not enough for the purposes at hand.

Here's a perfect thought experiment to help you understand the problem. Get out a piece of paper, and try, briefly, to sketch how a bicycle looks. It doesn't need to be a work of art; just try to place the seat, handles, tires, pedals, and bike chain in the right place. Can you do it?

Don't cheat by just trying to visualize the bicycle. Actually see if you can draw it. If you don't have a pencil or paper handy, you can simulate it by saying which things connect to what. Have you tried it?

Interestingly, Rebecca Lawson's study asked participants to do exactly this. As the illustrations clearly show, most participants had no idea how the machines were assembled, even though they used them all the time and believed they understood them quite well. The illusion of understanding is very often the barrier to deeper knowledge, because unless that competency is actually tested, it's easy to mislead yourself into thinking you understand more than you do. Feynman's and Einstein's approach to understanding propositions by demonstrating them prevents this problem in a way that's hard to do otherwise.

Were you one of the lucky ones who managed to put the chains on correctly? Try the exercise again, except this time with a can opener. Can you explain how it works? How many gears are there? How does it cut the lid open? This one is much harder, yet most of us would say we understand can openers!

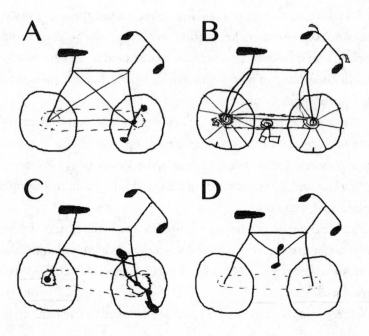

Rule 3: Always Start with a Concrete Example

Human beings don't learn things very well in the abstract. As the research on transfer demonstrates, most people learn abstract, general rules only after being exposed to many concrete examples. It's not possible to simply present a general principle and expect that you can apply it to concrete situations. As if presaging this observation, Feynman himself would supply concrete examples even when they were not given. Working through an explicit example in his mind's eye, he could follow along and see what the math was trying to demonstrate.

This process of following along with one's own example forces a deeper level of processing the material as it is being presented. A finding from the literature on memory, known as the levels-of-processing effect, suggests that it isn't simply how much time you spend paying attention to information that determines what you retain but, crucially, how you think about that information while you pay attention to it. In one study of this effect, participants were asked to review a list of words; half of them were told it would be for a test (and thus they were motivated to learn it), while the others were simply told to review the list. Within each group, participants were again split by what orienting technique they used to review the list. Half were asked to notice whether or not the words contained the letter *e*, a relatively shallow level of processing, while the others were asked if the word was pleasant or not, a deeper processing of the meaning of the word, not merely its spelling. The result was that motivation played no difference; telling students to study for a test didn't impact how much they retained. However, the orienting technique did make a large difference. Those who processed the words deeply remembered almost twice as much as those who simply scanned their spelling.

Feynman's habit of developing a concrete instance of a problem can be seen as an example of this deeper form of processing, which not only enhances later retention but also fosters an intuitive understanding. This technique also enables some feedback, because when it's not possible to imagine an appropriate example, that's evidence that you don't understand something well enough and would benefit from going back a few steps and learning the material better before continuing. Using feedback-rich processes to test whether or not he knew something was a hallmark of Feynman's learning style.

Rule 4: Don't Fool Yourself

"Don't fool yourself" was one of Feynman's most popular aphorisms, to which he added, "and you're the easiest person to fool." He was deeply skeptical of his own understanding. He presaged the current replication crisis in psychology, attacking what he perceived as many social scientists fooling themselves into believing they had discovered something they had not. I suspect that part of this insight arose from the fact that he had cultivated such rigorous standards for what he counted as knowing.

The Dunning-Kruger effect occurs when someone with inadequate understanding of a subject nonetheless believes he or she possesses more knowledge about the subject than the people who actually do. This can occur because when you lack knowledge about a subject, you also tend to lack the ability to assess your own abilities. It is true that the more you learn about a subject, the more questions arise. The reverse also seems to be true, in that the fewer questions you ask, the more likely you are to know less about the subject.

One way to avoid this problem of fooling yourself is simply to ask lots of questions. Feynman took this approach himself:

"Some people think in the beginning that I'm kind of slow and I don't understand the problem, because I ask a lot of these 'dumb' questions: 'Is a cathode plus or minus? Is an an-ion this way, or that way?'"* How many of us lack the confidence to ask "dumb" questions? Feynman knew he was smart and had no problem asking them. The irony is that by asking questions with seemingly obvious answers, he also noticed the not-so-obvious implications of the things he studied.

The opposite tendency, to avoid asking questions in the vain attempt to appear knowledgeable, has considerable costs. While lecturing in Brazil, Feynman's students would often complain when he asked simple questions that they knew the answers to already, instead of just lecturing. Why waste valuable classroom time on such exercises? The answer, Feynman eventually realized, was that they didn't know the answers but didn't want to admit it in front of everyone else in the class, wrongly assuming that they were the only ones who didn't know it. Explaining things clearly and asking "dumb" questions can keep you from fooling yourself into thinking you know something you don't.

THE FEYNMAN TECHNIQUE

When I first read about Feynman, I was inspired to try to formulate many of these different observations into a concrete method I could apply to my own studies. What resulted was something I

---

* Calling this the Feynman Technique was possibly unwise. It's unclear if Feynman ever used this exact method, so I may have inadvertently given the technique an illustrious history it doesn't possess. In addition, one of Feynman's great contributions to physics was in the form of "Feynman Diagrams." So, the Feynman Technique can lead to diagrams, although not necessarily Feynman Diagrams!

named the Feynman Technique and applied extensively during my MIT Challenge. The purpose of using this technique is to help develop intuition about the ideas you are learning. It can be used when you don't understand an idea at all or simply when you understand something a little but really want to turn it into a deep intuition.

The method is quite simple:

1. Write down the concept or problem you want to understand at the top of a piece of paper.
2. In the space below, explain the idea as if you had to teach it to someone else.
   a. If it's a concept, ask yourself how you would convey the idea to someone who has never heard of it before.
   b. If it's a problem, explain how to solve it and—crucially—why that solution procedure makes sense to you.
3. When you get stuck, meaning your understanding fails to provide a clear answer, go back to your book, notes, teacher, or reference material to find the answer.

The crux of this method is that it tries to dispel the illusion of explanatory depth. Since many of our understandings are never articulated, it's easy to think you understand something you don't. The Feynman Technique bypasses this problem by forcing you to articulate the idea you want to understand in detail. Just as drawing a bicycle quickly confirms whether you have a basic grasp of how it is put together, using this technique will quickly reveal how much you really understand of your subject. Now any gaps in your understanding will become obvious as you struggle to explain key parts of the idea.

The technique itself has some nuances and can be applied in a few different ways that might be helpful, depending on your specific intuitive deficit.

Application 1: For Things You Don't Understand at All

The first way to use this approach is when you don't understand something at all. In this case, the easiest way is to do it with the book in hand and go back and forth between your explanation and the one in the book. This lacks the benefits of retrieval practice, but it can often be essential when the explanation you've been given baffles you. Feynman himself did something similar when presented with what he saw to be philosophical gobbledygook:

> I had this uneasy feeling of "I'm not adequate," until finally I said to myself, "I'm going to stop, and read *one sentence* slowly, so I can figure out what the hell it means."
>
> So I stopped—at random—and read the next sentence very carefully. I can't remember it precisely, but it was very close to this: "The individual member of the social community often receives his information via visual, symbolic channels." I went back and forth over it, and translated. You know what it means? "People read."

Although Feynman's method was aimed more at illustrating the deliberately confusing nature of the prose rather than trying to understand a nuanced meaning, the same method can help whenever you're learning anything that goes over your head.

I used this technique when taking a class on machine vision during the MIT Challenge. I didn't understand photogrammetry, a technique of determining the 3D shape of an object based on a series of 2D pictures taken under different lighting conditions. It involved some tricky concepts, so I wasn't quite sure how it worked. With my textbook at my side, I wrote a few pages of notes, trying to sketch out the broad strokes of the idea so I could get the general gist of it.

Application 2: For Problems You Can't Seem to Solve

A second way to apply this is for solving a difficult problem or mastering a technique. In this instance, it's very important to go through the problem step by step alongside the explanation you generate, rather than simply summarizing it. Summarizing may end up skipping over the core difficulties of the problem. Going deeper may take time, but it can help you get a strong grasp over a new method in one go, rather than needing numerous repetitions to memorize the steps.

I applied this to a class in computer graphics for a technique I was struggling with called grid acceleration. This is a method of speeding up the performance of ray-traced rendering systems by avoiding analyzing objects that "obviously" won't be on the part of the screen you're drawing. To get a better handle on this, I walked through the problem with the technique, drawing a little snowman that I imagined rendering, with lines shooting out of an eyeball representing the camera.

Application 3: For Expanding Your Intuition

A final way to apply this method is to ideas that are so important that it would really help if you had a great intuition about them. In this application of the method, instead of focusing on explaining every detail or going along with the source material, you should try to focus on generating illustrative examples, analogies, or visualizations that would make the idea comprehensible to someone who has learned far less than you have. Imagine that instead of trying to teach the idea, you are being paid to write a magazine article explaining the idea. What visual intuitions would you use to pin down the abstractions? Which examples would flesh out a general principle? How could you make something confusing feel obvious?

I applied this to understanding the concept of voltage in an early class on electromagnetism during the MIT Challenge. Though I was comfortable using the concept in problems, I didn't feel that I had a good intuition of what it was. It's obviously not energy, electrons, or flows of things. Still, it was hard to get a mental image of an abstract concept on a wire. Going through this technique and comparing the equations to the ones for gravity, it's clear that voltage is to the electrical force as height is to the gravitational force. Now I could form a visual image. The wires were like troughs of water at different heights. Batteries were like pumps, moving the water up. Resistors were like hoses dropping down, of various widths to impede the flow of water draining down. Although this picture of troughs and hoses wasn't necessary for solving the equations, it stuck with me and helped me reason my way out of new situations more easily than if voltage had just been an abstract quantity.

DEMYSTIFYING INTUITION

When many people look at a genius like Richard Feynman, they're inclined to focus on his seemingly effortless intuitive leaps. In his playful style and rebellious impulses, he may seem to defy the stereotype that learning requires hard work. However, as we go beneath the surface, it becomes clear that he shared much in common with the other ultralearners I've studied. He worked hard on understanding things, and he put incredible amounts of his spare time into mastering the methods that made his intuition work. In his early days in college, he and a friend went back and forth over the early books on quantum mechanics, racing ahead of their classmates to understand it. He even made a meticulous timetable to allocate hours to his many intellectual pursuits. Even in his trivial obsessions, he displayed a streak for aggressive methods; while learning lock picking, for example, he trained himself to go through all the possible combinations, practicing them repeatedly: "I got it down to an absolute rhythm so I could try the 400 possible back numbers in less than half an hour. That meant I could open a safe in a maximum of eight hours—with an average time of four hours."

When people hear about geniuses, especially the iconoclastic ones such as Feynman, there's a tendency to focus on their gifts and not their efforts. I have no doubt that Feynman possessed gifts. But perhaps his greatest one was his ability to merge tenacious practice and play. He approached picking locks with the same enthusiasm for solving puzzles that he did for unraveling the secrets of quantum electrodynamics. It's this spirit of playful exploration that I want to turn to in the final principle of ultralearning: experimentation.

CHAPTER XII

PRINCIPLE 9

Experimentation

Explore Outside Your Comfort Zone

Results? Why, I have gotten lots of results! I
know several thousand things that won't work.
—*Thomas Edison*

If you were to read his story without seeing his art, Vincent van
Gogh would be the last person you would expect to become one
of the most famous painters of all time. He started at a late age,
twenty-six. Art is a field of precocity, and famous masters typically display their gifts early. Pablo Picasso's cubist style, for
example, came out of his already being able to paint realistically
as a child, allowing him to boldly declare that it had taken him
"four years to paint as Rafael, but a lifetime to paint like a child."
Leonardo da Vinci was apprenticed as a painter in his teens. One
story has him, as a youth, painting a monster on a peasant's shield
only to have it resold to the duke of Milan. Salvador Dalí had his

first exhibit before his fourteenth birthday, already showcasing the talent that would make him famous. Van Gogh, in contrast, was delayed and did not possess any obvious signs of ability. It was only after failing as an art dealer and minister that he picked up the brush. An art seller and family friend, H. G. Tersteeg, believed his artistic aspirations were put on to mask his laziness. "You started too late," he declared. "Of one thing I am sure, you are no artist. . . . This painting of yours will be like all the other things you started, it will come to nothing."

Worse than the fact that he started late, however, was that van Gogh simply wasn't very good at drawing. His drafting was crude and childlike. When he finally convinced models to sit for portraits—no small task in light of the Dutchman's famously difficult personality—it took him many attempts to get anything resembling a likeness. While in a brief stint at a Parisian atelier, he even studied next to future leaders of the Post-Impressionist movement, such as Henri de Toulouse-Lautrec. However, un-like the effortless quality with which Toulouse-Lautrec cap-tured the likeness of a scene with a few flicks of his wrist, van Gogh struggled. "We considered his work too unskillful," one classmate recalled. "His drawings had nothing remarkable about them." In the end, his inability to fit in with his classmates, lack of talent, and off-putting manners had him leaving the studio after less than three months.

His late start and lack of obvious talent were compounded by his temperament. Nearly everyone who entered his life would eventually reject him, as his manic enthusiasm and fraternal sol-idarity would inevitably sour into bitter fights with nearly ev-ery person he met. Near the end of his life, he was routinely placed in mental asylums, varyingly diagnosed with disorders from "acute mania with generalized delirium" to "a kind of epilepsy." His outbursts, or "attacks," as he referred to them, alienated him from people who could potentially serve as his peers, mentors, and teachers. As a result, despite having attempted formal schooling, van Gogh was largely self-taught, capturing only brief moments of more traditional education in the moments during which he could hold on to friendships before pushing people away.

It was van Gogh's mysterious and untimely death that cut short the artistic career that was so late to begin. At thirty-seven, he died of a bullet wound to the stomach. Although his death was suspected to be a suicide, his biographers Steven Naifeh and Gregory White Smith consider accident or foul play more likely; he may possibly even have been shot by one of the village youths who played pranks on him and called him the *fou roux*, or "crazed redhead."

In spite of all this, van Gogh has become one of the most famous painters of all time. *The Starry Night, Irises,* and *Vase with Fifteen Sunflowers* have become icons. On four separate occasions, a van Gogh piece has become the most expensive painting ever sold, including his *Portrait of Dr. Gachet*, which was sold for more than $82 million. Van Gogh's signature swirls of color, thick impasto application, and strong outlines have made many people consider his paintings some of the greatest of all time.

How can we explain these discrepancies? How does someone who starts late, with no obvious talent and many handicaps, nonetheless become one of the world's greatest artists, with one of the most recognizable and distinctive styles? To understand van Gogh, I want to turn to the ninth and final principle of ultralearning: experimentation.

HOW VAN GOGH LEARNED TO PAINT

Put yourself in van Gogh's shoes for a moment. You've failed miserably as an art dealer, despite your family connections. You've failed as a preacher. Now you're embarking on a new profession—painting—even though you have difficulties in drawing things accurately. What would you do? Van Gogh's response to this challenge was a pattern that would repeat throughout his life. First, he would identify a learning resource, method, or style and pursue it with incredible vigor, creating dozens, if not hundreds, of works in that direction. After this burst of intensity, aware of his still-existing deficiencies, he would apply himself to a new resource, method, or style and start again. Although there's no evidence van Gogh thought of the connection, I see a parallel between this pattern and the one used by successful scientists: hypothesis, experiment, results, repeat. Perhaps inadvertently, van Gogh's aggressive, experimental strides into painting allowed him to mature into not merely a proficient painter but an unforgettably unique one.

Van Gogh's experimentation began when he was first trying to become an artist. The normal route to an artistic career in those days was to attend an art school or apprentice in a studio. Van Gogh, due to the fact that others did not see him as possessing much talent and his odd temperament, did not have much luck with those traditional avenues. Therefore he turned to self-education, pursuing home-study courses that promised to teach him the basics of drawing. In particular, he made heavy use of Charles Bargue's *Exercices au fusain* (Charcoal Exercises) and *Cours de dessin* (Drawing Course), as well as Armand Cassagne's

Guide de l'alphabet du dessin (Guide to the ABCs of Drawing). They were thick books with graduated exercises on which aspiring artists could work step by step to improve their drawing skill. According to his biographers, van Gogh "devoured these big books . . . page by page, over and over." Van Gogh himself reported to his brother, Theo, "I have now finished all sixty sheets," adding "I worked almost a whole fortnight, from early morning until night." Copying was another strategy van Gogh employed early on that he would continue late into his artistic career. Jean-François Millet's *The Sower* was one of his favorite pictures to copy, which he did again and again. He also applied himself to sketching from life early, in particular models for portraits, which he struggled with due to his difficulties with drafting accurately.

Van Gogh studied from other artists, friends, and mentors. Anthon van Rappard convinced him to try out reed pen and ink and adopt the mature artist's style of short and fast strokes. Another artist, Anton Mauve, persuaded him to try a variety of different media: charcoal and chalk, watercolor, and Conté crayon. Often those attempts were not successful. During their stay together at the house where van Gogh would later cut off his ear, Paul Gauguin pushed the Dutchman to paint from memory, mute his colors, and adopt new materials for different effects. Those tactics didn't work for van Gogh, whose weaknesses in drafting were exacerbated by not having the scene directly in front of him, and the different materials went against the style that would later make him famous. Experiments, however, needn't always be successful to have value, and van Gogh had many opportunities for trying new techniques.

Van Gogh experimented not just with materials and methods

but also with the philosophies that underpinned his art. Although he is most famous for strong, vibrant colors, that wasn't his initial intention. Originally, he leaned toward the profundity of muted, grayer tones, as witnessed in an early work *The Potato Eaters.* "Scarcely any color is not gray," he argued. "In nature one really sees nothing else but those tones or shades." He was fully convinced of that and based his work on it accordingly. However, he would later switch to the exact opposite: bright, complementary colors, often imposed on a scene rather than being brought out from nature. His stance on contemporary artistic movements flitted about; at first he preferred traditional painting to the new Impressionist style, and later he shifted to the avant-garde, opting for bold forms rather than verisimilitude.

There are two important things to note about van Gogh's experiments in art. The first is the variety of methods, ideas, and resources he applied. Since he struggled with many aspects of painting, I believe that variation was important to his eventually finding a style that would work for him—one that would take advantage of his strengths and diminish the significance of his weaknesses. Although virtuoso talents might be able to latch on to the first style of instruction they are presented with and follow it to completion, others require a great deal of experimentation before the right method sticks. The second important thing to note is his intensity. Like all the ultralearners I've discussed so far, van Gogh was tenacious in his efforts to become an artist. Despite receiving much negative feedback and discouragement, he pursued his art relentlessly, sometimes producing as much as a new painting every day. These two factors, variation and aggressive exploration, enabled him to push through his early obstacles and produce some of the most iconic and brilliant works ever painted.

EXPERIMENTATION IS THE KEY TO MASTERY

When starting to learn a new skill, often it's sufficient simply to follow the example of someone who is further along than you. In discussing the principles of ultralearning, metalearning comes first. Understanding how a subject breaks down into different elements and seeing how others have learned it previously, thus providing an advantageous starting point. However, as your skill develops, it's often no longer enough to simply follow the examples of others; you need to experiment and find your own path.

Part of the reason for this is that the early part of learning a skill tends to be the best trodden and supported, as everyone begins at the same place. As your skills develop, however, not only are there fewer people who can teach you and fewer students you could have as peers (thus lowering the total market for books, classes, and instructors), but you also start to diverge from those you're learning from. Whereas two complete novices have quite similar knowledge and skills, two experts might have quite different sets of skills that they've already acquired, thus making improving those skills an increasingly personalized and idiosyncratic adventure.

A second reason for the value of experimentation as you approach mastery is that abilities are more likely to stagnate after you've mastered the basics. Learning in the early phases of a skill is an act of accumulation. You acquire new facts, knowledge, and skills to handle problems you didn't know how to solve before. Getting better, however, increasingly becomes an act of unlearning; not only must you learn to solve problems you couldn't before, you must unlearn stale and ineffective approaches for solving those problems. The difference between a novice programmer and a master isn't usually that the novice cannot solve certain problems. Rather, it's that the master knows the best way to solve a problem, which will be the most efficient and clean and cause the fewest headaches later on. As mastery becomes a process of unlearning over accumulation, experimentation becomes synonymous with learning as you force yourself to go outside your comfort zone and try new things.

A final reason for the increasing importance of experimentation as you approach mastery is that many skills reward not only proficiency but originality. A great mathematician is one who can solve problems others cannot, not merely a person who can solve previously solved problems easily. Successful business leaders are those who can spot opportunities others cannot, not merely those who can copy the style and strategy of those before them. In art, it was not only van Gogh's skill but his originality that made him one of the most celebrated painters to have ever lived. As creativity becomes valuable, experimentation becomes essential.

THREE TYPES OF EXPERIMENTATION

In experimentation you can see different levels play out, both in van Gogh's path as an artist and as a model for your own explorations:

1. Experimenting with Learning Resources

The first place to experiment is with the methods, materials, and resources you use to learn. Van Gogh did this extensively at the beginning of his artistic career, trying out different artistic media, materials, and learning techniques: following home-study courses, watching fellow artists, sketching from life and in the studio, and more. This kind of experimentation is useful in helping you discover the guides and resources that work best for you. It's important, however, that your impulse to experiment be matched with a drive to do the necessary work. Although van Gogh tried many different approaches when he first started teaching himself to draw and paint, he also produced an enormous quantity of work based on each of those methods.

A good strategy to take is to pick a resource (maybe a book, class, or method of learning) and apply it rigorously for a predetermined period of time. Once you apply yourself aggressively to that new method, you can step back and evaluate how well it is working and whether you feel it makes sense to continue with that approach or try another.

2. Experimenting with Technique

In the beginning, experimentation tends to focus on materials. However, in most domains of learning, the options for what to learn next expand faster and faster, so the question becomes not "How can I learn this?" but "What should I learn next?" Languages are a prime example. The same basic set of vocabulary and phrases dominates most beginner resources. As you improve, however, the amount of things you could possibly learn next becomes larger and larger. Should you learn to read literature? Converse fluently on a professional topic? Read comic books? Have business discussions? The specialized vocabulary, phrases, and cultural knowledge in each field multiply, so it becomes necessary to choose what to master.

Once again, experimentation plays a pivotal role. Pick some subtopic within the skill you're trying to cultivate, spend some time learning it aggressively, and then evaluate your progress.

Should you continue in that direction or pick another? There's no "right" answer here, but there are answers that will be more useful to the specific skill you're trying to master.

3. Experimenting with Style

After you've matured in your learning a bit, the difficulty often switches from which resources to learn from or which techniques you'd like to master to the style you'd like to cultivate. Although there are some skills that have one and only one "correct" way of doing things, this is not true of most. Writing, design, leadership, music, art, and research all involve developing certain styles, which have different trade-offs. Once you master the basics, there is no longer one "right" way to do everything but many different possibilities, all of which have different strengths and weaknesses. This affords another opportunity for experimentation. Van Gogh tried out many different styles for producing art, varying from those of traditional painters such as Millet, to Japanese woodblock prints, and studied the techniques used by his artist friends such as Gauguin and Rappard. There is no one correct answer, although, like van Gogh, you may find that certain styles work better than others with your unique combination of strengths and weaknesses.

The key to experimenting with different styles is to be aware of all the different styles that exist. Once again van Gogh provides a good model, as he spent an enormous amount of time studying and discussing the works of other artists. That gave him a large library of possible styles and ideas he could adapt to his own work. Similarly, you might want to identify masters in your own line of study and dissect what makes their styles successful to see what you can emulate or integrate into your own approach.

In each level of experimentation, the choices broaden and the possible options to explore go up exponentially. There's a tension, therefore, between spending time trying out different resources, techniques, and styles, and concentrating your efforts on a single approach long enough to become proficient at it. This tension often resolves itself as you cycle through exploring a new avenue in learning and then buckling down to learn it deeply before moving on to something else. Whatever else his failings, it was this pattern of trying out an idea and working on it aggressively that van Gogh applied brilliantly.

THE MINDSET OF EXPERIMENTATION

There are parallels between the mindset required to experiment and what the Stanford psychologist Carol Dweck refers to as growth mindset. In her research, she distinguishes between two different ways of looking at one's own learning and potential. In a fixed mindset, learners believe that their traits are fixed or innate and thus there's no point in trying to improve them. In a growth mindset, in contrast, learners see their own capacity for learning as something that can be actively improved. In some ways, these two types of mindsets become self-fulfilling prophecies. Those who think they can improve and grow, do; those who think they are fixed and immutable remain stuck.

The parallel with the mindset required for experimentation is clear. Experimenting is based on the belief that improvements are possible in how you approach your work. If you think your learning styles are fixed or that you have certain immutable strengths and weaknesses that will keep you from trying out different ways to approach your skills, you won't be able to experiment at all.

I see the experimental mindset as an extension of the growth mindset: whereas the growth mindset encourages you to see opportunities and potential for improvement, experimentation enacts a plan to reach those improvements. The experimental mindset doesn't just assume that growth is possible but creates an active strategy for exploring all the possible ways to reach it. To get into the right mental space for experimenting, you need not only to see your abilities as something you can improve but understand that there is a huge number of potential avenues to do this. Exploration, not dogmatism, is the key to realizing that potential.

HOW TO EXPERIMENT

Experimenting sounds simple but can be quite tricky to implement in practice. The reason is that a flurry of random activity doesn't usually translate into mastery. In order to work, experimenting requires understanding what learning problems you're facing and coming up with possible ways to resolve them. Here are a few tactics that can help you integrate experimentation into your ultralearning projects.

Tactic 1: Copy, Then Create

This is the first strategy for experimenting, which we can see in van Gogh's work. Though van Gogh is best known for his original pieces, he also spent a lot of time copying drawings and paintings he liked by other artists. Copying simplifies the problem of experimentation somewhat because it gives you a starting point for making decisions. If you're learning to paint, as van Gogh did, the possibilities of what kinds of art you can create and techniques you can apply are so vast that it can be difficult or impossible to decide among them. However, if you start by emulating another artist, you can use that foothold to venture further in your own creative directions.

This strategy has another advantage beyond simplifying the choices available to you. In attempting to emulate or copy an example you appreciate, you must deconstruct it to understand why it works. This can often highlight things that the other person does exceptionally well that weren't obvious from the beginning. It may also dispel illusions you may have had about an aspect of the work you thought was important but upon emulating the other person's work you realize was not.

Tactic 2: Compare Methods Side-by-Side

The scientific method works by carefully controlling conditions so that the difference between two situations is limited to the variable being studied. You can apply this same process to your experiments in learning by trying two different approaches and varying only a single condition to see what the impact is. By applying two different approaches side by side, you can often quickly get information not only about what works best but about which methods are better suited to your personal style.

I applied this to learning French vocabulary. I wasn't sure how effective mnemonics would be so, for a month, I would find a list of fifty new words every day, put together from my regular reading or random encounters with the language, and for half I would simply look them over with their translations I got from the dictionary. With the other half, I made an effort to use a visual mnemonic to link the two meanings. Then I compared how many of the words I remembered from each list on a later test, with words picked randomly from each side. The result is something you would probably expect after reading the chapters on retrieval and retention: I remembered the words I used mnemonics for at almost twice the rate of those I didn't. That showed that even if creating the mnemonics took a bit more time, they were worth it.

There are two advantages to doing split tests. The first is that as in scientific experiments, you will get much better information about which method works best if you limit the variation to only the factor you want to test. The second is that by solving a problem multiple ways or applying multiple solution styles to it, you will increase your breadth of expertise. Forcing yourself to try different approaches encourages experimentation outside your comfort zone.

Tactic 3: Introduce New Constraints

The challenge of learning in the beginning is that you don't know what to do. The challenge of learning in the end is that you think you already know what to do. It's this latter difficulty that causes us to rerun old routines and old ways of solving problems that are encouraged through habit, not always because the old way is actually best. A powerful technique for pushing out of those grooves of routine is by introducing new constraints that make the old methods impossible to use.

It's practically an axiom of design that the best innovations come from working within constraints. Give a designer unlimited freedom, and the solution is usually a mess. On the other hand, creating specific constraints in how you can proceed encourages you to explore options that are less familiar to you and sharpens your underlying skills. How can you add limitations to force yourself to develop new capacities?

Tactic 4: Find Your Superpower in the Hybrid of Unrelated Skills

The traditional path to mastery is to take a well-defined skill and practice it relentlessly until you have become insanely good at it. This is the path taken by many athletes, who train for decades to perfect their shot, jump, kick, or throw. However, for many areas of creative or professional skills, another, more accessible, path is to combine two skills that don't necessarily overlap to bring about a distinct advantage that those who specialize in only one of those skills do not have. For instance, you might be an engineer who becomes really good at public speaking. You may not be the best possible engineer or the best possible presenter, but combining those two skills could make you the best person to present on engineering topics for your company at conferences, thus giving you access to new professional opportunities. Scott Adams, the creator of *Dilbert*, likened his own success to following this strategy by combining his background as an engineer with an MBA and a cartoonist.

This level of experimentation often plays out over multiple ultralearning projects. After I completed my MIT Challenge, I could apply the programming knowledge I had obtained to write scripts to automatically generate flash cards for learning Chinese. Such synergies become possible once you start exploring how one skill you've already acquired can impact another.

Tactic 5: Explore the Extremes

Van Gogh's art pushed well outside normal conventions along many dimensions. His thick application of paint was far away from the thin layers of glazes used by Renaissance masters. His quick application was far more rapid than the careful brushstrokes of other painters. His colors were bold, often garish, instead of subtle. If you were to draw a chart that mapped out van Gogh's style compared with those of other painters, you would probably see that he lay along the extreme in many dimensions.

An interesting result from mathematics is that as you get to higher and higher dimensions, most of the volume of a higher-dimensional sphere lies near its surface. For instance, in two dimensions (a circle), just under 20 percent of its mass lies in the outer shell described by a tenth of the radius. In three dimensions (a sphere), that number rises to almost 30 percent. In ten dimensions, almost three-quarters of the mass is in that outermost layer. You can imagine learning a complex subject as akin to trying to find an optimal point in a region of higher-dimensional space—except that instead of length, width, and height, those dimensions might be the qualitative dimensions of work, such as van Gogh's complementarity of colors, application of paint, or some other aspect of skill that can be applied in some varying degree of intensity. What this means is that the more complicated a domain of skill is (i.e., the more dimensions it contains), the more space will be taken up by applications of that skill that are extreme across at least one of those dimensions. This suggests that for many skills, the best option is going to be extreme in some way, since so many more of the possibilities are themselves extreme. Sticking to the middle and playing it safe isn't the correct approach because that allows you to explore only a small subset of the total possibilities for your work.

Pushing out to an extreme in some aspect of the skill you're cultivating, even if you eventually decide to pull it back to something more moderate, is often a good exploration strategy. This allows you to search the space of possibilities more effectively, while also giving you a broader range of experience.

EXPERIMENTATION AND UNCERTAINTY

Learning is a process of experimenting in two ways. First, the act of learning itself is a kind of trial and error. Practicing directly, getting feedback, and trying to summon up the right answers to problems are all ways of adjusting the knowledge and skills you have in your head to the real world. Second, the act of experimenting also lies in the process of trying out your learning methods. Try out different approaches, and use the ones that work best for you. The principles I've tried to articulate in this book should provide good starting points. But they are guidelines, not iron rules; starting points, not destinations. Only by experimenting will you be able to find the right trade-offs between different principles—for instance, when directness is more important and when you should focus on drills or whether retention or intuition is the main obstacle to learning. Experimenting will also help you decide among small differences in approach that no list of principles could possibly cover exhaustively.

Having a mindset of experimentation will also encourage you to explore beyond what you feel most comfortable doing. Many people stick to the same routines, the same narrow set of methods, they apply to learning everything. As a result, there are a lot of things they struggle to learn because they don't know the best way to do so. Copying exemplars, running tests, and pushing to extremes are all ways to push outside your ingrained habits and try out something different. That process will teach you not just abstract learning principles but concrete tactics that will accommodate your personality, interests, strengths, and weaknesses. Are you better off learning a language through practicing speaking or engaging in lots of input through movies and books? Are you better off learning programming by building your own game or working on open-source projects? These questions don't have a single correct answer, and people have achieved success using a wide variety of different methods.

My own experience with learning has been one of constant experimentation. In university, I focused a lot on making associations and connections. During the MIT Challenge, I switched to making practice the foundation. In my first experience learning a language, I was sloppy, speaking English most of the time. In the second round, I experimented with going to another extreme, to see if I could avoid that sticking point. While doing projects, I've had to adjust my methods frequently. Even though it was only thirty days long, my portrait-drawing challenge involved a lot of trial and error from starting by doing sketches and, when my progress using that approach slowed, trying to do sketches even faster to get more feedback. When that, too, had reached its limits, I spent some time learning a different technique altogether to achieve greater accuracy.

Embedded in my successes are many failures—times where I thought I could get something to work and it ended up failing miserably. Early on in learning Chinese, I thought I could use some kind of mnemonic system to remember the words, with colors for tones and memorized symbols for the syllables. That was because my normal sounds-like method for visual mnemonics wasn't working with the words, which all sounded so different from English. The result was a total failure, and it didn't work at all! Other times, my experiments with new methods worked out great. Most of the techniques I've shared in this book thus far started as ideas I wasn't sure would pan out.

Experimentation is the principle that ties all the others together. Not only does it make you try new things and think hard about how to solve specific learning challenges, it also encourages you to be ruthless in discarding methods that don't work. Careful experimentation not only brings out your best potential, it also eliminates bad habits and superstitions by putting them to the test of real-world results.

CHAPTER XIII

Your First Ultralearning Project

The beginning is always today.
—*Mary Shelley*

By now you're probably eager to start your own ultralearning project. What things could you learn that you have put off due to fears of inadequacy, frustration, or lack of time? What old skills could you take to new heights? The biggest obstacle to ultralearning is simply that most people don't care enough about their own self-education to get started. As you've read this far, I doubt that is true of you. Learning, in whatever forms it takes, is something that's important to you. The question is whether that spark of interest will ignite into a flame or be smothered prematurely.

Ultralearning projects aren't easy. They require planning, time, and effort. Yet the rewards are worth the effort. Being able to learn hard things quickly and effectively is a powerful skill. One successful project tends to lead to others. It's usually the first project that requires the most thought and care. A solid, well-researched, well-executed plan can give you the confidence to face harder challenges in the future. A bungled attempt is not a disaster, but it may make you reluctant to pursue future projects of a similar nature. In this chapter, I'd like to tell you everything I've learned about how to get it right.

STEP 1: DO YOUR RESEARCH

The first step in any project is to do the metalearning research required to give you a good starting point. Planning ahead will avoid a lot of problems and prevent you from having to make drastic changes to your learning plan before you've even started making progress. Research is a bit like packing a suitcase for a long voyage. You may not bring the right items, or you may forget something and need to buy it on the road. However, thinking ahead and packing your bags correctly will prevent a lot of fumbling later. Your ultralearning "packing" checklist should include, at a minimum:

1. **What topic you're going to learn and its approximate scope.**
   Obviously, no learning project can begin unless you figure out what you want to learn. In some cases, this is obvious. In others, you may need to do further research to identify which skill or knowledge would be most valuable. If your goal is to learn something instrumentally (to start a business, get a promotion, do research for an article), learning what you need to learn is important and will suggest how wide and deep you need to go. I suggest starting with rather a narrow scope, which can expand as you proceed. "Learning enough Mandarin Chinese to hold a fifteen-minute conversation on

simple topics" is a lot more constrained than "Learn Chinese," which may include reading, writing, studying history, and more.

2. **The primary resources you're going to use.** This includes books, videos, classes, tutorials, guides, and even people who will serve as mentors, coaches, and peers. This is where you decide what your starting point will be. Examples: "I'm going to read and complete the exercises in a book on Python programming for beginners" or "I'm going to learn Spanish through online tutoring via italki.com" or "I'm going to practice drawing by making sketches." In some subjects, static materials will determine how you proceed. In others, they will be supports to back up your practice. In any case, they should be identified, purchased, borrowed, or enrolled in before you begin.

3. **A benchmark for how others have successfully learned this skill or subject.** Almost any popular skill has online forums where those who have learned the skill previously can share their approaches. You should identify the things other people who have learned the skill have done to learn it. This doesn't mean you need to follow exactly in their footsteps, but it will prevent you from completely missing something important. The Expert Interview Method in chapter 4 provides a good method for following up on this.

4. **Direct practice activities.** Every skill and subject you're learning will be used somewhere eventually, even if it's as simple as using it to learn something else. Thinking about how you might use the skill can enable you to start finding opportunities to practice it as early as possible. If direct practice is impossible, you should nonetheless identify opportunities for practice that mimic the mental requirements of using the skill.

5. **Backup materials and drills.** In addition to the principal

materials and methods you'll be using, it's a good idea to look at possible drills and backup materials you may want to use. Backup materials are often good if you recognize that a certain tool or set of materials might be useful but you don't want to be overwhelmed in the beginning.

STEP 2: SCHEDULE YOUR TIME

Your ultralearning project doesn't need to be an intensive, full-time endeavor to succeed. However, it will require some time investment, and it's better to decide on how much time you are willing to devote to learning in advance than simply hope that you'll find the time later. There are two good reasons for planning your schedule ahead of time. The first is that this way you subconsciously prioritize your project by setting it down on your calendar ahead of other things. The second is that learning is often frustrating and it is almost always easier to click over to Facebook, Twitter, or Netflix. If you don't set aside time to learn, it will be a lot harder to summon up the motivation to do so.

The first decision you should make is how much time you're going to commit. This is often dictated by your schedule. You may have a gap in employment that allows intensive learning, but only for a month. Alternatively, you may have a full schedule that permits you to devote only a few hours per week to learning something new. Whatever time you can commit, decide on it in advance.

The second decision you need to make is when you are going to learn. During a few hours on Sunday? By waking up an hour earlier and putting in the time before work? In the evening?

During lunch breaks? Once again, the best thing is to do whatever makes it easiest based on your schedule. I recommend setting a consistent schedule that is the same every week, rather than trying to fit in learning when you can. Consistency breeds good habits, reducing the effort required to study. If you have absolutely no choice, an ad hoc schedule is better than none, but it will require more discipline to sustain.

If you do have some flexibility in your schedule, you may want to optimize it. Shorter, spaced time chunks are better for memory than crammed chunks are. However, some types of tasks, such as writing and programming, have a long warm-up time that may benefit from longer uninterrupted time chunks. The best way to find out what is best for you is to practice; if you find it takes a long time to warm up, opt for longer spaces in your schedule. If you find you can get to work within a few minutes of starting, shorter chunks of time spread out will be helpful for long-term retention.

The third decision you need to make is the length of time for your project. I generally prefer shorter commitments to longer ones because they are easier to stick with. An intensive project that lasts a month has fewer potential interruptions from life or from your motivation changing and waning. If you have a big goal you want to accomplish that can't be done in a short time frame, I suggest breaking it up into multiple smaller ones of a few months each.

Finally, take all this information and put it into your calendar. Scheduling all the hours of work on the project in advance has important logistical and psychological benefits. Logistically, this will help you spot potential conflicts in your schedule due to vacations, work, or family events. Psychologically, it will help you remember and act on your initial plan better than if it were written on a piece of paper tucked into a desk drawer. What's more, the act of scheduling demonstrates your seriousness about doing the project.

I can clearly remember writing down my hypothetical studying schedule before starting the MIT Challenge. It had me up and studying by 7 a.m. and working until 6 p.m., with only a short break for lunch. Although my actual schedule, in practice, rarely reached that ideal (even in my most intensive early days, I almost never got in eleven hours straight), the mere act of writing down the schedule helped prepare me psychologically for the project ahead. If you're unwilling to put time into your calendar, you're almost certainly unwilling to put in time to study. If you're waffling at this stage, that's a good sign your heart isn't really in the right place to get started.

As a bonus step, for those who are embarking on longer projects of six months or more, I strongly recommend doing a pilot week of your schedule. This is simple: test your schedule for one week before you commit to it. This will give you firsthand knowledge of how difficult it will be and prevents overconfidence. If you already feel burned out after the first week, you may need to make adjustments. There's no shame in going back and retooling your plan to make it fit your life better. Making this kind of adjustment is a lot better than giving up midway because your plan was doomed from the start.

STEP 3: EXECUTE YOUR PLAN

Whatever plan you started with, now is the time to do it. No plan is perfect, and you may realize that what you're doing for learning departs from the ideal, as established by the ultralearning principles. You may notice that your plan relies too much on passive reading rather than retrieval practice. You may see that the way you're practicing is a winding detour away from where you will actually want to use it. You may feel as though you're forgetting things or memorizing them without really understanding them. That's okay. In some cases, you won't be able to have the perfect learning approach because the resources to do so don't exist. However, becoming sensitive to how the way you're learning isn't aligned with the principles is a good way to feel out changes you can make to improve it.

Here are some questions to ask yourself to determine whether you're slipping from the ideal:

1. **Metalearning.** Have I done research into what are the typical ways of learning this subject or skill? Have I interviewed successful learners to see what resources and advice they can recommend? Have I spent about 10 percent of the total time on preparing my project?

2. **Focus.** Am I focused when I spend time learning, or am I multitasking and distracted? Am I skipping learning sessions or procrastinating? When I start a session, how long does it take before I'm in a good flow? How long can I sustain that focus before my mind starts to wander? How sharp is my attention? Should it be more concentrated for intensity or more diffuse for creativity?

3. **Directness.** Am I learning the skill in the way I'll eventually be using it? If not, what mental processes are missing from my practice that exist in the real environment? How can I practice transferring the knowledge I learn from my book/class/video to real life?

4. **Drill.** Am I spending time focusing on the weakest points of my performance? What is the rate-limiting step that is holding me back? Does it feel as though my learning is slowing down and that there's too many components of the skill to master? If so, how can I split apart a complex skill to work on smaller, more manageable components of it?

5. **Retrieval.** Am I spending most of my time reading and reviewing, or am I solving problems and recalling things from memory without looking at my notes? Do I have some way of testing myself, or do I just assume I'll remember? Can I successfully explain what I learned yesterday, last week, a year ago? How do I know if I can?

6. **Feedback.** Am I getting honest feedback about my performance early on, or am I trying to dodge the punches and avoid criticism? Do I know what I'm learning well and what I'm not? Am I using feedback correctly, or am I overreacting to noisy data?

7. **Retention.** Do I have a plan in place to remember what I'm learning long term? Am I spacing my exposure to information so it will stick longer? Am I turning factual knowledge into procedures that I'll retain? Am I overlearning the most critical aspects of the skill?

8. **Intuition.** Do I deeply understand the things I'm learning, or am I just memorizing? Could I teach the ideas and procedures I'm studying to someone else? Is it clear to me why what I'm learning is true, or does it all seem arbitrary and unrelated?

9. **Experimentation.** Am I getting stuck with my current resources and techniques? Do I need to branch out and try new approaches to reach my goal? How can I go beyond mastering the basics and create a unique style to solve problems creatively and do things others haven't explored before?

Together these principles serve as directions, not destinations. In each case, look at how you're currently progressing through your materials, and see what you could do differently. Do you need to switch resources? Do you need to stick to the same resources but spend more time on a different kind of practice? Should you seek out new environments for feedback, directness, or immersion? These are all subtle adjustments you can make along the way.

STEP 4: REVIEW YOUR RESULTS

After your project is finished (or if you end up putting it on pause for some reason), you should spend a little time analyzing it. What went right? What went wrong? What should you do next time to avoid making those same mistakes?

Not all of your projects will be successful. I've had ultralearning projects that I felt good about. I've had others that didn't work out as well as I had hoped. Although the tendency is to blame willpower and motivation, very often the problems with projects can be traced back to their conception. I worked on one project devoted to improving my Korean, after my trip, by investing five hours per week. It wasn't as successful as I had hoped because I didn't invest enough time in focusing on immersive, direct practice from the get-go. Instead, my study method depended a lot on textbook exercises, which were boring and didn't transfer too well to the real world. If I had thought a little more about it, I would have spent a week or two ahead of time trying to find places to practice, instead of trying to pivot midway, when I was already losing some motivation. This struggle illustrates that mastering the principles is a lifelong process. Even after many experiences learning languages, and knowing what works well, I slipped into a less effective approach because I didn't plan my project adequately. In other cases a project might not work out as you had hoped, but that lesson will still be valuable. I started with a project to learn cognitive science more deeply, going from a book list. Eventually, however, a lot of that project morphed into a desire to do research for this book, which put me into contact with a lot of science, now combined with an outlet for a more direct way to apply it.

Even your successful projects are worth analyzing. They can often tell you more than your failures because the reasons a successful project succeeded are the very elements you want to retain and replicate for the future. With ultralearning, as with all self-education, the goal isn't merely to learn one skill or subject but to hone and enhance your overall learning process. Each successful project can be refined and improved for the next one.

STEP 5: CHOOSE TO MAINTAIN OR MASTER WHAT YOU'VE LEARNED

After you've learned your skill and analyzed your efforts, you have a choice to make. What do you want to do with the skill? With no plan in place, most knowledge eventually decays. This can be alleviated somewhat by following the principles of ultra-learning. However, all knowledge decays without any form of intervention, so the best time to make a choice about how you're going to handle that is right after you learn something.

Option 1: Maintenance

The first option is to invest enough practice to sustain the skill but without any concrete goal of getting it to a new level. This can often be accomplished by setting up a habit of regular practice, even if it is a minimal one. As mentioned in the chapter on retention, one of the worries I had after the year without English project was that learning languages so intensively over a short period of time might lead not just to rapid learning but to rapid forgetting. As a result, I made an effort to continue practice after the trip finished, spending thirty minutes a week on each language in the first year and thirty minutes a month on each language in the year after that.

Another option is to try to integrate the skill into your life. This is how I maintain my programming skills, where I write Python scripts to handle work tasks that would otherwise be cumbersome or annoying. This kind of practice is more sporadic, but it ensures that I will keep it up enough to make it usable. This kind of lightweight usage is far from the deep math and algorithms I learned from my MIT coursework, but it is enough to keep a foot in the door if I want to embark on a bigger project at a later time.

Forgetting, as was discovered by Hermann Ebbinghaus more than a hundred years ago, falls off with an exponentially decaying curve. That means that memories that are retained for longer are less and less likely to be forgotten when you follow up at a later date. This pattern suggests that maintenance practice, too, can fall off on a decaying rate, so that the bulk of the knowledge you've acquired will be preserved. This means you might want to start with a habit of more serious practice but reduce the time spent on it a year or two after your project is finished to still get most of the benefit, as I did with the languages I studied.

Option 2: Relearning

Forgetting isn't ideal, but for many skills the costs of relearning the skill later are smaller than the costs of keeping it continuously sharp. There are a couple reasons for this. First, you may have learned more than you actually need, so if some of that knowledge selectively decays due to disuse, it is automatically going to be the less important knowledge that you acquired. I studied a lot of MIT subjects that I don't think I'll ever use again, although understanding the gist of them might come in handy later. Therefore, keeping my ability to prove theorems of modal logic, for example, up to date has only marginal value. Knowing what modal logic is and where I might apply it in case I want to learn something that requires it is probably enough.

Relearning is generally easier than first-time learning. Although performance on tests drops off dramatically, the knowledge is likely inaccessible rather than completely forgotten. This means that doing a refresher course or practice series can be enough to reactivate most of it in a fraction of the time it took to learn it initially. This may be the optimal strategy for subjects that you need to use infrequently and for which situations for using them won't pop up without warning. Often, recognizing that a certain domain of knowledge is helpful for a particular problem type is more important than the details of solving the problem, since the latter can be relearned but forgetting the former will cut you off from solving those problems.

Option 3: Mastery

The third option, of course, is to dive deeper into the skill you have learned. This can be done through continued practice at a lighter pace or by following up with another ultralearning project. A common pattern I've noticed in my own learning is that an initial project covers a wider territory and some basics and exposes new avenues for learning that were previously obscured. You might identify a subtopic or branch of skill within the domain you were learning before that you want to follow up. Otherwise, you may decide to transfer a skill learned in one place to a new domain. One of my goals after returning from my trip to China was to learn to read Chinese better, which had been only an incidental goal while I was traveling there.

Mastery is a long road that extends far beyond a single project. Sometimes the barriers you overcome in your initial effort are enough to clear the way for a slow process of accumulation to reach eventual mastery. In many domains, getting started is quite frustrating, so it's difficult to practice without a certain amount of effort. After that threshold is reached, however, the process switches to being one of accumulating huge swaths of knowledge and therefore can proceed at a more patient pace. On the other hand, some projects will get stuck, and you will need to spend time unlearning and push through your frustrations again to get ahead. Those kinds of projects benefit more from the precise and aggressive methods of ultralearning to reach eventual mastery.

ALTERNATIVES TO ULTRALEARNING: LOW-INTENSITY HABITS AND FORMAL INSTRUCTION

At the start of this book, I pointed out that ultralearning is a strategy. Being a strategy implies that it is good for solving certain problems. Given that the practice is somewhat uncommon, I wanted to spend the book focused on this strategy, rather than try to give a diffuse description of all possible ways you can learn effectively. However, now that I've done that, I think it's worthwhile to touch on two other strategies that can work with ultralearning, in different contexts.

None of the ultralearners I encountered approaches learning the same way for every kind of learning they do. Benny Lewis, for example, does do intensive learning projects for languages, but he has learned most of his languages over repeat visits to the countries they are spoken in, digging deeper into languages he previously established in intense bursts. Roger Craig did learn aggressively to win at *Jeopardy!*, but he also engaged in more leisurely acquisition of trivia when his appearance on the game show wasn't imminent. Being an ultralearner doesn't imply that everything one learns has to be done in the most aggressive and dramatic fashion possible. I want to briefly consider the two main alternative strategies to ultralearning to see how they fit into a bigger picture of lifelong learning.

Alternative Strategy 1: Low-Intensity Habits

Low-intensity habits work well when engaging in learning is spontaneous, your frustration level is low, and learning is automatically rewarding. In these cases, when the barriers to learning are fairly low, all you need to do is show up. No fancy project, principles, or effort is required. Once you reach a conversational level in a language, for instance, it's often fairly easy to travel and live in a country where it is spoken, accumulating more and more vocabulary and knowledge over a longer period of time. Similarly, once you become good enough at programming to use it for your job, the job itself will push you to learn new things at a regular pace. If you've mastered the basics of a subject so that you can read denser books about it, reading books on the topic is mostly a matter of putting in time, not developing ingenious learning strategies.

Of course, there's a spectrum of habits, from zero-effort, spontaneous engagement to the high-effort, rapid skill acquisition of ultralearning. Most habits are somewhere in between, necessitating a bit of effort but perhaps not the full-scale intensity of an ultralearning project. You may have learned enough Excel to create your own spreadsheet macros, but you don't always find opportunities or time to use it, so you need to push yourself a little to practice. You may have learned public speaking well, but it still takes some guts to go onstage. The decision of whether the right step forward is to set up long-term habits or to create a concentrated ultralearning project is often not crystal clear and may depend more on your personality and life constraints than a hard-and-fast rule.

Habits tend to work best when the act of learning is mostly a process of accumulation, adding new skills and knowledge. Ultralearning and more deliberate efforts are better suited to when improvement in a field requires unlearning ineffective behaviors or skills. Increasing your vocabulary in a foreign language is often a slow process of accumulation; you are learning words you didn't know before. Improving your pronunciation, on the other hand, is an act of unlearning. You're training yourself to use different muscular movements that aren't natural to you. Ultralearning also tends to be better for areas in which learning has greater frustration barriers and psychological obstacles that make any form of practice too great an effort to be an easily established habit.

Throughout this book, we've explored the trade-off that occurs between doing what's effective for learning and what's easy and enjoyable. Sometimes what's the most fun isn't very effective, and what's effective isn't easy. This trade-off may push you to opt for easier, more enjoyable forms of learning that sacrifice some effectiveness. However, in my own experience, I've noticed that enjoyment tends to come from being good at things. Once you feel competent in a skill, it starts to get a lot more fun. Therefore, although a tension between the two can exist in the short term, I think pursuing aggressive ultralearning projects is often the surer way to enjoy learning more, as you're more likely to reach a level where learning automatically becomes fun.

Alternative Strategy 2: Formal, Structured Education

In the beginning of this book, I explained that ultralearning is self-directed, although not necessarily solitary. Being self-directed is about who is making decisions, not about whether other people are involved. Therefore, there is no contradiction in pursuing ultralearning within a school or university. That might be the best way to learn the skills you want to acquire. Just treat it like any other resource.

That distinction notwithstanding, I think it's worth talking about other reasons you might want to pursue formal education rather than ultralearning. The most obvious is to acquire credentials. If those are necessary or recommended for your chosen line of work, you might need to be satisfied with making sacrifices for your learning in order to acquire them. The message of this book isn't that you should drop out of school to learn on your own but that you should take control over your own learning, wherever that may be. Another reason to pursue formal education is that it creates a learning environment that may be beneficial. Although many aspects of school are woefully indirect and ineffective, others fare much better. Design and art schools often function like apprenticeships. Some programs allow for team projects that are difficult to start on your own. Finally, the postgraduate levels of academia create communities where immersion is possible, so that you acquire not only the ideas that are written down in books and papers but those that are communicated indirectly between experts in their fields. Ultralearning isn't a rejection of those opportunities, and I would be disappointed if I were to be misread as arguing that they don't exist or are better replaced with a solitary learning effort. The correct mindset to cultivate isn't a rejection of anything slower or standardized but a recognition that the possibilities for learning anything are considerably broader than they might first appear.

LIFELONG LEARNING

The goal of ultralearning is to expand the opportunities available to you, not narrow them. It is to create new avenues for learning and to push yourself to pursue them aggressively rather than timidly waiting by the sidelines. This is not going to be a method suitable for everyone, but for those who feel inspired to use it, I hope it provides a start.

CHAPTER XIV

An Unconventional Education

> Give me a dozen healthy infants, well-formed,
> and my own specified world to bring them
> up in and I'll guarantee to take any one at
> random and train him to become any type of
> specialist I might select—doctor, lawyer, artist,
> merchant-chief and, yes, even beggar-man
> and thief.
> —*Psychologist John Watson*

Judit Polgár is widely considered to be the best female chess player of all time. At age seven, she won her first game against a chess master while blindfolded. At twelve, she was ranked fifty-fifth of all chess players in the world by the Fédération Internationale des Échecs (FIDE) (World Chess Federation). By fifteen, she had become the youngest-ever grand master, beating the illustrious Bobby Fischer's previous record by one month. At her peak, Polgár was ranked eighth in the world and competed in the World Chess Championship, the only woman ever to have done so.

Chess is a game dominated by adult men. Thus a young girl competing was bound to arouse both the curiosity and the prejudice of her competitors. Grand master Edmar Mednis, facing off against the young Polgár, noted that he was very careful to play his best against the young prodigy, noting that "Grandmasters don't like to lose to 10-year-old girls, because then we make the front page of all the papers." Some of her competitors celebrated the obvious genius of Polgár's play. Grand master Nigel Short said Polgár might be one of the "three or four great chess prodigies in history." Mikhail Tal, a former world champion, suggested when Polgár was still twelve that she might eventually be a contender for world champion.

Garry Kasparov was less convinced. The former world champion has been considered by many to be the best chess player of all time. He is most famous for his matches against IBM's Deep Blue chess computer, winning against the machine in 1996 and losing in 1997, marking the transition to machine dominance in a game that had historically been considered one of the highest expressions of human creativity and intelligence. Kasparov was less than enthusiastic about the young Polgár's chances. "She has fantastic chess talent, but she is, after all, a woman. It all comes down to the imperfections of the feminine psyche. No woman can sustain a prolonged battle."

That casual prejudice erupted into a full-blown controversy during their first match. Polgár, then only seventeen, sat across from the chess legend and former world champion, playing in a tournament in Linares, Spain. Although chess is often viewed as being coldly rational, as both players calculate moves with precision to reach their final outcome, the psychological effect of sitting opposite the dominant Russian can't be understated. Given that incredible tension, it was almost unbelievable when Kasparov, on move thirty-four, placed his knight and, after briefly lifting his fingers from the piece, changed his mind and moved it to a better square. Polgár was stunned. According to chess rules, once a player stops touching a piece, the move is done; no changes are allowed. Half in disbelief, she glanced at the referee, expecting him to indicate that Kasparov had cheated. Yet the referee didn't challenge the grand master. Reeling from the move, Polgár lost the game.

Asked why she hadn't challenged the illegal move herself, Polgár explained, "I was playing the World Champion and didn't want to cause unpleasantness during my first invitation to such an important event. I was also afraid that if my complaint was overruled I would be penalized on the clock when we were in time pressure." Still, after the game had concluded, she was incensed. She confronted Kasparov later in the hotel bar, demanding "How could you do this to me?" "She publicly accused me of cheating," Kasparov said when defending himself against the accusation. "I think a girl of her age should be taught some good manners." It would be years before the two spoke again, but whereas Kasparov was already well established in the chess world, Polgár was just getting started.

Polgár is singular, not only for her prowess in the male-dominated game but also for how she learned to play. Unlike other famous players, such as Bobby Fischer, who spontaneously developed an obsession for the game, Polgár's chess genius was no accident. Instead, it started with one man's mission to raise genius children.

THE MAKING OF A GENIUS

Years before Judit faced down the legendary grand master, before her meteoric rise in chess or even her first game, her father, László Polgár, had made a decision: he was going to raise a genius. While studying intelligence in college, he had begun contemplating his project before having children or even a wife. "A genius is not born but educated and trained," he maintained. Studying the biographies of hundreds of great intellectuals, he was convinced that genius could be manufactured. "[W]hen I looked at the stories of geniuses," he later remarked, "I found the same thing. . . . They all started at a young age and studied intensively."

But first, he had to find a partner for his pedagogical experiment. He found it in Klára, a Ukrainian foreign-language teacher. Far from your normal love letters, he first corresponded to her by explaining his idea to raise genius children. After agreeing to his proposal, the two met and were married in the Soviet Union before moving back to his native Hungary. Together the couple had three children, Zsuzsa, Zsófia, and Judit. Although Judit ended up becoming the most intensely competitive and famous, all three became world-class chess players, with Zsuzsa also becoming a grand master and Zsófia reaching the status of international master.

Living modestly in a cramped apartment, László and his wife decided to devote themselves full-time to their project of raising genius children. Their strategy was to begin the girls' educations early, at age three, and move on to specialization in one domain no later than six. They would start by introducing the girls to that subject slowly, in short bursts, and turn it into play so that they would actively want to practice rather than feel coerced into doing so. That strategy, however, didn't specify a topic. László and Klára considered many different possible study topics for their daughters, from foreign languages to mathematics. Eventually they settled on chess, because it was objective and progress in it was easy to measure. No doubt the preeminence of chess intellectually in the socialist countries in those years added weight to their decision to focus on the game.

Despite the emphasis on chess, László did not believe that such specialization needed to come at the sacrifice of the broader education of his girls. All three learned foreign languages (Zsuzsa, the eldest, learned eight), as well as mathematics, table tennis, swimming, and other subjects. The decision to focus on chess with all three girls was made for a practical reason: given the intense devotion both parents would need to have, in both resources and time, spreading their effort over three different disciplines would have been more than their budget or schedule could bear.

Zsuzsa was first to start. She began playing at age four. Six months later, she was going with her father to the smoke-filled chess clubs of Budapest, playing against elderly men—and winning. By the time it was Judit's turn, she was already motivated to begin her training. Zsuzsa and Zsófia played chess in a small room László had devoted to the game, and she didn't want to be left out.

Soon the three girls were a team, traveling around to compete against much older, usually male players. Their shared mission created camaraderie rather than jealousy in playing a game that very often pushed back against their unusual status. The Hungarian Chess Federation's policy was for women to compete in women-only chess events. However, László was strongly against the idea. "Women are able to achieve results similar, in fields of intellectual activities, to that of men," he felt. "Chess is a form of intellectual activity, so this applies to chess. Accordingly, we reject any kind of discrimination in this respect." Discrimination had already prevented Zsuzsa from obtaining the grand master title at fifteen. As the youngest, by the time Judit approached some of these barriers, her elder sisters had already broken them down a bit, allowing her to forgo competing in the women-only tournaments.

Although their education was the same and all three reached impressive heights in chess, the girls' prowess was not identical. Zsófia was the weakest of the three; although reaching the impressive grade of international master, she later decided to retire from chess to focus on art and her family. Zsuzsa had specialized a little less in chess from an early age, learning eight languages, which her father admits might have distracted her from reaching her maximum chess potential. Judit started more slowly, according to Zsuzsa, but she had the strongest work ethic, being "obsessed" with chess to a degree unusual even for her family.

REMATCH WITH KASPAROV

Eight years after her controversial defeat by Kasparov, Judit had another opportunity to face down the legendary grand master. In the 2002 Russia Versus the Rest of the World match in Moscow, Judit faced off against Kasparov in a game of rapid chess, with only twenty-five minutes allocated to each player. Judit played the Ruy Lopez, or Spanish game, named after the sixteenth-century Spanish bishop and chess strategist. This uses one of the most common openings in chess, moving the knight and bishop out to good squares on the second and third moves. Kasparov countered with the Berlin Defense, moving a second knight to the board while ignoring the potential danger created by his opponent's bishop; this was known to be a solid counter, often leading to draws. Kasparov wasn't taking any chances. After a flurry of exchanged pieces, both positions were quite close. Judit, as white, had castled her king into safety. Kasparov, as black, in losing the opportunity for the same safety had retained both the light-square and dark-square bishop pair, a powerful combination that can often be decisive in winning a game. Judit calmly pushed forward, cornering one of Kasparov's bishops and neutralizing his advantage. Slowly but surely, her moves continued to improve her position while Kasparov's looked increasingly questionable. Eventually, the small positional advantages Judit had accumulated over the midgame were threatening to win. Two pawns down and facing looming threats of checkmate, Kasparov resigned.

Following his defeat, Kasparov revised his earlier assessment of Judit's ability and indeed the idea of women competing against men at the highest levels of the game. "The Polgárs showed that there are no inherent limitations to their aptitude—an idea that many male players refused to accept until they had unceremoniously been crushed by a twelve-year-old with a ponytail."

THE AFTERMATH OF AN EXPERIMENT

László Polgár's conviction that he could turn any healthy child into a genius is the kind of assertion that would make him seem like a crackpot if he hadn't actually been successful. Nonetheless, a careful reader will note that, as far as experiments go,

László's has plenty of holes that keep it from becoming a model of scientific purity. There was no control group, for starters. All three of the Polgár sisters received the same education. There was no fourth sister who went to school the normal way and missed László's special training. There was no randomization. László didn't adopt a random child to raise in his unusual system but taught his own. This also means that the influence of genetics can't be ignored. The success of all three children may be owing to hereditary, rather than acquired, talents. There was also no blinding. All the Polgárs knew they were part of something special, a unique mission that set them apart from other families. Therefore, in the ongoing debate between nature and nurture, the success of the Polgár sisters might be suggestive of the role that unorthodox education can play, but it is far from definitive.

Despite its failings as a purely scientific experiment, the Polgárs' work is certainly a window into what might be possible. All three girls achieved enormous results in chess. Although we cannot know for certain, it seems likely that they could have succeeded in any number of other domains as well. Similarly, although László's methods were strange, it does not seem as though the girls suffered, either in their broader education or in their emotional well-being. They were self-confident and happy and grew up to be successful, emotionally stable adults with loving families of their own. When asked whether his strange pedagogical methods had robbed his girls of a normal childhood, László argued the opposite, that it is a normal, mediocre education that often leads to unhappiness. Interviewer Endre Farkas, who worked with Polgár on his book *Raise a Genius!*, asked, "Have they been educated too narrowly, deprived of a carefree childhood? I was able to observe the Polgárs. . . . One can clearly see that they are happy."

EDUCATING ULTRALEARNERS?

Prior to doing the research for this book, all of the ultralearners I had met were ambitious self-starters. I was convinced that ultralearning was something that held great potential for the individual. However, owing to the intensity and commitment required by the learners themselves, I was skeptical that ultralearning would have any direct implications for the educational system at large. Children already struggle against onerous studying conditions, and it seemed to me that increasing the intensity of study would only increase their stress and anxiety.

Psychologists recognize a large difference between goals that people pursue intrinsically, based on their own interests, decisions, and targets, and goals that they pursue extrinsically, pushed by overbearing parents, punishing curricula, or demanding employers. The latter type, because the motivation for conforming to them comes primarily from outside social pressures, are the cause of much misery. Tales of depression, anxiety, and even suicide are distressingly common in environments where the pressure to perform on standardized tests is ratcheted up to an uncomfortable degree. Ultralearning, because it is a self-driven quest, not an externally imposed obligation, doesn't need to be this way. However, because of its very nature, it was unclear to me whether it was something that could be taught.

The Polgár sisters are an unusual case, then, because although they were trained from a very early age and worked incredibly hard, they don't seem to have been harmed psychologically by the pressure. In contrast to the stereotypical "tiger" parent, their parents encouraged their unusual specialization through play and positive feedback, not authority and punishment. The

Polgár sisters all continued to play chess well into adulthood at the competitive level. The obsession into chess, therefore, was clearly something fostered, rather than imposed. At the same time, their participation in the experiment wasn't exactly voluntary. László had dreamed up his goal of raising geniuses before knowing whether his children would consent to the program, so it was not a case of each daughter discovering for herself that she wanted to devote herself to an intense regimen of chess practice. That feature of the Polgár experiment interested me the most, because it seemed as if László and Klára had found a loophole in the normal expectation that pushing others to study intensely necessarily leads to misery.

HOW TO RAISE AN ULTRALEARNER

László Polgár wrote a book entitled *Raise a Genius!*, documenting his unorthodox approach to education.* In the book, he outlined his strategy for turning any normal, healthy child into a genius, provided the parents are willing to go to the extremes he and his wife dedicated to the task.

The first step is to start early. The child's education should begin no later than three, and specialization should begin no later than six. Although the exact extent to which learning becomes harder as we age is unclear, there is evidence from fields such as music and languages that children's brains are more plastic and flexible when younger. László took this idea to the

---

* This book, *Raise a Genius!*, originally appeared under the title *Neveli zsenit!* I'm indebted to the blogger Scott Alexander and his readers for sourcing a translation in English.

extreme, encouraging training far earlier than is typical for early-childhood education.

The second step is to specialize. Although the Polgár sisters did learn languages, mathematics, sports, and other subjects, their focus was always on chess. László noted that "starting from 4–5 they played chess 5 or 6 hours a day." That specialization seems to have had two roles in his strategy for raising geniuses. The first is that it took advantage of any hypothetical flexibility that younger children possess to learn new subjects easily. The second is that by specializing in one subject, the children could reach proficiency at a much younger age. Winning against older and more experienced opponents in chess built their confidence and competitive spirit, so that they actively wanted to practice more to improve themselves. Had the girls spread their intellectual appetites too broadly, they might not have developed the self-confidence that leads to intense practice.

The third step was to make practice into play. Chess, being a game, is naturally suited to play. However, László was insistent on introducing all subjects to the girls as forms of play. When the girls would become distracted or get up and wander around while playing a game, they weren't punished but encouraged to let their minds wander as they sought a solution. Keeping the game fun and light, especially when the children were young, was a key stepping-stone to developing the drive and self-confidence that would support more serious efforts later. However, it is important to keep in mind as László insisted that "play is not the opposite of work" and "a child does not need play separate from work, but meaningful action," adding "learning presents them with more enjoyment than a sterile game." Play and work combined in the Polgárs' approach to learning, with no rigid boundary between them.

Fourth, László was careful to create positive reinforcement to make chess a pleasant, rather than frustrating, experience. "Failure, suffering, and fearfulness decrease achievement. Following a number of successive failures, even a damaging inhibitory complex can be created," he explained. Starting with the behaviorists, psychologists have known in detail that having a positive experience, such as winning a game, can create a desire to repeat the behaviors that led to the experience. Negative experiences, from losing, becoming confused, or experiencing frustration against a stronger opponent to facing an opponent who is too easy and robs the player of the satisfaction of pulling off an impressive victory, reduce enthusiasm. László carefully established the cycles of positive feedback early on. In the beginning, when he was still a stronger player than the girls, he would adjust his play so that they would be challenged but still win enough of the time to find the game enjoyable. "We should make sure not to always win against the child; we should let them win sometimes so that they feel they are also capable of thinking," he wrote, adding "At the start it is most important to awake interest. . . . We should get the child to love what they do—to such a degree that they do it almost obsessively."

Finally, László was entirely against coercing learning. Self-discipline, motivation, and commitment, he felt, must come from the girls themselves. He explained, "One thing is certain: one can never achieve serious pedagogical results, especially at a high level, through coercion." He also felt that "one of the most important educational tasks is to teach self-education." This final step of his process was particularly important for his daughters, as they quickly outpaced their father's ability. Had they not been encouraged to develop their own abilities to teach themselves and adjust their learning, they might have become decent chess players but certainly not grand masters.

In addition to these basic principles, László and Klára devoted themselves intensely to providing every opportunity for the advancement of their daughters, cultivating a database of more than two hundred thousand matches, buying every chess textbook they could find, and recruiting chess tutors for their daughters. The girls lacked no opportunity to study and improve at the game. The Polgár house, with diagrams of chess positions hanging on the walls, became a temple devoted to the practice of the ancient game. For László and Klára, raising their children was more than a full-time job as they fostered the girls' talents by both assembling resources and schooling them at home.

ULTRALEARNING PRINCIPLES IN ACTION

In addition to Polgár's principles of raising genius children, I found it interesting that all of the principles of ultralearning I've discussed this far were present in their approach to learning.

1. Metalearning
Polgár devoted himself full-time to understanding how people learn chess and under what conditions his daughters would thrive. He created a huge library of chess positions, strategies, and lists of games, which, in the days before widespread internet, was no small feat. He also articulated a plan for coaching the girls in the game when they were still very young, starting by teaching them first to name the squares on the board and later to know how the pieces moved. That slow progression allowed

the girls to pick up the game even before their other cognitive abilities had developed.

2. Focus

László considered "the ability to handle monotony, the capability to sustain interest and persistent attention" as key traits he aimed to instill in his daughters. The girls twice took part in twenty-four-hour chess marathons when they were fifteen, nine, and eight years old, respectively, needing to complete one hundred matches in the time period. Chess is a game not merely of flickers of brilliance but of endurance and stamina. Training focus was a large part of László's system for his daughters, as he encouraged them to focus their minds on the problem and not get distracted.

3. Directness

László took his daughters to games with men when they were as young as four, showing them how the game was to be played against opponents who would really offer a challenge. The girls played many, many games of chess, which formed the backbone of their abilities. That allowed them to learn not only how to play the game well but also to deal with variables such as time pressure and the psychological insecurities of playing against older, more intimidating opponents. By using chess timers even for casual games, the girls practiced in an environment that more closely matched the one they would face in tournaments.

4. Drill

László varied the approaches to studying the game, starting his daughters off with memorizing first the names of the squares, then the movements of key pieces. Chess puzzles, hanging from the walls of the house, became the girls' homework, as they had to solve the different tactical positions and come up with creative solutions. Blitz and blindfolded games allowed the girls to get better at thinking more quickly and mentally simulating the game.

5. Retrieval

For retrieval, László explained, "We should not tell them everything; we should try to get the child to say something!" Using what he described as the "Socratic method" for chess, posing questions his girls must answer instead of telling them to remember a presolved solution, he was using the right method to encourage the expansion of their memory and understanding. Blindfolded games, once again, formed a powerful component of the girls' strategy. By practicing without looking at the board, it forced them to cultivate the ability to follow positions in their head, which was useful not only for retaining key chess patterns long term but also for honing the ability to simulate moves on the board that an opponent might play.

6. Feedback

László encouraged considerable play with real opponents but was careful to select "suitable partners, who have a generally similar playing ability." Interestingly, the feedback here was carefully controlled, not only to provide the girls with enough challenge (the Polgárs' insistence on playing in men's tournaments in order to face such a challenge was an example of this) but also to avoid too great a challenge when their abilities were still nascent. Cultivating positive feedback was important early on, and László was ever ready to adjust the flow of the game to make sure it was at a level that would stimulate further play.

7. Retention

László focused on having the girls recall chess patterns from memory and increased the speed of games to make elements of their play more automatic and less susceptible to forgetting. Memorizing chess patterns is a large part of playing the game successfully, and this was aided both by spaced practice and through specialized drills such as blitz and blindfolded games.

8. Intuition

Mirroring the Feynman Technique, László encouraged his girls to write articles about chess, explaining, "If one writes an article, one considers a matter more deeply than without a goal, thinking alone or speaking with someone about it." The girls were also encouraged to come up with creative solutions to problems. Play, not merely in the sense of chess being a game but also in the sense of an unconstructed, goalless activity, was part of the teaching strategy. Coming up with interesting solutions and challenging the girls to think of tricks and new insights allowed them to explore outside what could be offered by memorizing past results.

9. Experimentation

As the Polgár sisters eclipsed their father in chess ability, their impetus to continue mastering the game increasingly had to come from within themselves. Each of the girls had to cultivate her own unique style and approach. Judit chose to focus on tricks and tactics, writing that "opening preparation was not at all important at that time. This may be a reason why, even today, my strongest area remains the middlegame." The girls' varying choices show that chess, like any creative skill, involves not merely a mastery of patterns but also choices about what skills and styles to cultivate within a vast range of possibilities.

Finally, the Polgárs embodied the idea of ultralearning at its broadest, with László arguing, "In my opinion, we should disseminate the idea of intensive learning in every field." The success of the Polgárs follows the same pattern as that of most of the ultralearners I have met: aggressive, enthusiastic self-education following the key principles of learning.

FOSTERING ULTRALEARNING IN THE HOME, SCHOOL, AND WORKPLACE

How can you foster ultralearning as a parent or educator or in an organization? Is it possible to help others self-confidently tackle difficult learning projects of their own design? Can you teach students not just the material they need to study but how to learn on their own, so they will be self-sufficient outside the classroom? Can you lead the individuals in your organization to learn more aggressively, filling gaps in their competency and achieving their full potential? These are all intriguing questions to which we don't yet have definitive answers.

In reading the scientific literature on learning and following up with the ultralearners' stories, I was struck not only by how much is known about learning already but by how many open questions exist for which researchers and autodidacts are still hazarding hypotheses. The complications expand exponentially once you introduce the social environment as well. Now it's no longer a question simply of individual cognition but of the emotions, culture, and relationships that start to influence learning in complex and unexpected ways. From this perspective, therefore, I'd like to cautiously suggest some starting points for fostering an environment that will support ultralearning, at home, in the school, or in the workplace. These suggestions aren't rules, but they can be seen as starting points for enabling others to capture the ultralearning spirit.

Suggestion 1: Create an Inspiring Goal

Better yet, allow people to design their own learning goals that inspire them. Inspiration is an essential starting point in the process of ultralearning. There must be something very compelling for a person to summon up the energy and self-discipline needed to learn. Sometimes that is the promise of a new skill bringing career opportunities. Coding boot camps, which have sprung up in the wake of high-paying programming jobs, push students through at a brutal pace, sometimes approaching eighty hours a week. The goal, however, is compelling enough to justify this investment: complete a rigorous program over the course of a few weeks, and you can rise up the ladder of high-salaried tech jobs in Silicon Valley and elsewhere. The process is intense, but the motive is compelling.

In other cases, the motivation for ultralearning comes from an intrinsic interest that becomes amplified. My own MIT Challenge started with feeling I had missed out by not studying computer science in school. Normally, that wouldn't have led to any large, structured effort to learn a lot of computer science. It was only when the idea of doing an entire degree in a shortened period of time, along with the research that made me think it might be possible, that my initial interest become a passionate commitment. Roger Craig, with his *Jeopardy!* exploits, was always interested in trivia competitions. It was only when he recognized that there might be a chance to appear on the famous television show that his interest became an obsession. Eric Barone took his love of a childhood video game and expanded it into an effort to create a better version. Seeking out people's natural interests for ultralearning means encouraging the sparks that already exist, rather than merely imposing on them the topics you feel would be most beneficial. Once people see the structure of an ultralearning project, they can start thinking for themselves what would be most interesting, exciting, and useful for them to work on. Tristan de Montebello started with the idea of ultralearning and only later chose to craft a public speaking project around it.

Suggestion 2: Be Careful with Competition

The Polgárs' example clearly indicates that early self-confidence can create an enthusiasm that leads to continued investment. You don't need to feel as though you're good at something to invest energy into learning. After all, becoming good at something is what learning is. However, you need to feel that you *could* be good at it. People tend to make their perceptions of inadequacy into immutable destinies: "I'm no good at math," "I can't draw anything but stick figures," "I don't have the language gene." Although there likely do exist real differences in innate ability, so that these pronouncements aren't completely false, they tend to ignore an important factor: motivation. When you see yourself as lacking the potential to be good at something or believe that you'll always be behind everyone else no matter how hard you work, it robs you of the motivation to work hard. Thus, although there are differences in ability among all of us, they can often be exacerbated by the affective dimension they create in how we feel about learning. Feel as though you're lousy at doing something, and you're robbed of the motivation to change.

The reference group you compare yourself to can have a powerful influence. I find it interesting that many, but not all, ultralearners aimed at projects that were so unusual that they made it hard to compare them to a normal reference group. De Montebello's public speaking competition certainly pitted himself against excellent public speakers. That might have created a feeling of inferiority, except that de Montebello could always explain to himself any perceived deficits as coming from trying such an ambitious project with so little prior experience. Had it been the case that instead of an individual ultralearning project, he had been pitted against a dozen other competitors with exactly the same prior experience, he might instead have rationalized any perceived inadequacy as his simply not being good enough. This suggests that the competitiveness of the project cuts both ways: When you have a natural talent and thus perform much better than the easily identifiable reference group, you'll have more motivation to practice and learn with intensity. However, if you come up short, it may rob you of the motivation to practice. The Polgárs used competition to their advantage. Because the girls' training started unusually young, they were always seen as precocious and the competitive environment enhanced their motivation. Had they started late or been put into a school where they weren't guaranteed to be star performers, their motivation might have been sapped.

To me, such motivational effects, coming from implicit comparison to a reference group, suggest adopting a twofold approach. If a person in whom you want to encourage an ultralearning spirit has a natural aptitude, competition is probably good. Seeing him-

or herself do well in direct comparison with others may encourage a commitment to further improvement. For a person who either is of moderate ability or is behind other people, such as learning a skill in a domain in which he or she has no experience, or who is starting to learn a new skill later in life, you should make an effort to make the project unique. This will encourage the person to frame his or her progress by comparing to his or her past self, not due to competition with others. Sometimes a project can start out being unique, thus sheltered from the harsh light of unfavorable comparison, and move to a more competitive environment once confidence has been established. For example, you might start learning programming by creating a game that is hard to compare to others but enter coding competitions as you begin to feel more competent.

Suggestion 3: Make Learning a Priority

Outside school, learning is usually seen as a by-product of doing a job, not the core goal. Though organizations often give lip service to ongoing training and education, it's usually in the form of workshops or seminars that one sits through passively before getting back to the real job at hand. Ultralearning, by encouraging direct, intensive practice, provides the opportunity for a kind of fusion project—one that accomplishes real objectives but is also designed to teach something new.

The normal protocol for assigning a project is to find the best person for the job and give them the task. A learning-driven approach would suggest instead that people who are not yet capable of doing the task might be assigned to the project. An ultralearning-driven work environment might consist of employees' spending a majority of their time on projects that are within or near their competency levels but devoting a specific fraction of their time to working on projects that are a leap above their current abilities. Although this is purely hypothetical, I imagine two benefits of this approach. First, it will create a culture of learning within an organization where people are always willing to try to solve problems they don't yet know how to solve, instead of expecting someone else to know the answer already. Second, it will help reveal talent by giving people challenges they can rise to. If mentorship opportunities and difficult projects are assigned only on a whim by managers, they will likely miss a lot of people who may have the ability to succeed in difficult positions but are never given the opportunity to do so.

At the highest levels, an ultralearning-driven culture also allows learning to go into areas where perhaps nobody else has a particular skill. Although going between established levels of skill is important, it is when one learns to do something that nobody else can do that learning becomes truly valuable.

CONCLUSION

In many ways, writing this book has been an ultralearning project. Although a writer researching for a book is hardly unique, not all ultralearning projects need to be one of a kind to matter to the person doing them. Sitting in my den at home are stacks of binders filled with thousands of pages of printed journal articles. My bookshelf now has dozens of obscure, out-of-print monographs on thin slices of the question of how people learn. Recordings of calls with various researchers helped me realize how much nuance there is to even simple questions such as "Is feedback helpful?" and "Why do people forget?" I've poured over numerous biographies of famous intellectuals, entrepreneurs, and scientists to try to arrive at an understanding of how they approached learning. In many ways, the process of writing this book was a reflection of its subject—an ultralearning project to write a book about ultralearning. Although I had a strong interest in the subject of learning and had browsed textbooks, articles, and biographies before I began research into this book, it was only after I started this structured project that I really began to dig deep.

Beyond research, this book was a challenge for me as a writer. My writing experience comes from blogging, not authoring books. Striking the right tone in a book is hard, and it's quite different from the casual daily missives in a blog. I knew from the start that I wanted to share the stories of others and their exploits, not just recount my own experiences. That was initially quite challenging. Most biographies and published stories don't focus on learning methods. Even when learning is the central theme of the story, most biographers are satisfied to be in awe of talent, rather than dig into the specific details of how a person did a particular thing. My research efforts frequently involved scouring a five-hundred-page biography for the several paragraphs in which concrete details about learning methods were mentioned in passing. Although this created challenges, it also forced me to develop new skills as a writer. I had to improve my research and writing skills in ways that more than a decade of penning blog articles never had. Even the style of the book created a skill-challenging project for myself. I'll leave it to you, the reader, to judge whether I was successful.

The metaproject of ultralearning to write a book about ultralearning also illustrates some important ideas. For one, although I've made enormous improvements in my writing ability and knowledge of cognitive science and stories of famous learning exploits, there is still far more to learn. Digging into the science, for instance, one can quickly develop a sense of vertigo standing atop the mountain of papers, theories, ideas, and experiments, all loosely connected to the topic of learning. Similarly, for every biography I read, there were hundreds I could not. For every ultralearning story I encountered, there were likely dozens more my searches didn't reveal. It is a profound error to claim that learning is about replacing ignorance with understanding. Knowledge expands, but so does ignorance, as with a greater understanding of a subject also comes a greater appreciation for all the questions that remain unanswered.

In the face of this, one must simultaneously have confidence and deep humility. Without the belief that progress in one's own knowledge and skill is possible, one cannot undertake the project required to generate it. This kind of confidence may be mistaken for arrogance by outsiders, as it can seem that an effort to learn something quickly and intensely is somehow an assertion that the subject is trivial or that, having learned something, one has learned everything. Instead, this confidence must be paired with deep humility. In every project I've undertaken, including this one, my thoughts upon concluding it were not to think I had finished but to suddenly become aware of how much further I could have gone. Before I started my MIT Challenge, I imagined that covering an undergraduate degree's worth of computer science concepts would be plenty. After I had finished, I could see how each topic I had learned could be multiplied into a doctorate's worth of research or a lifetime spent coding to fully understand it. My experience in learning languages to a level where I could hold conversations made me realize how many more words, expressions, nuances of culture, and difficult communication situations were left to explore. Finishing a project, therefore, isn't usually accompanied by a sense of finishing learning but by the creation of a feeling of possibility as one's eyes are opened to all the things left to learn.

It's this aspect of learning that I find most interesting. Many pursuits in life have a kind of saturation point, after which the longing for more of a thing eventually diminishes as you get more of it. A hungry person can eat only so much food. A lonely person can have only so much companionship. Curiosity doesn't work this way. The more one learns, the greater the craving to learn more. The better one gets, the more one recognizes how much better one could become. If you finish reading this book and have been encouraged to try your own project, this would be my greatest hope—not that you'd be successful at your project but that your ending would be a beginning. That by opening a small crack in all the possibly knowable things there are in the world, you might peer through and find there is far, far more than you had ever imagined.

ACKNOWLEDGMENTS

This book could not have happened without the help, advice, and work provided by many different people. First, I would like to thank Calvin Newport. Had it not been for his early encouragement, I might never have pursued writing a book about this topic. I would also like to thank Benny Lewis, whose early inspiration and endless advice over the years has had such a strong influence on my thoughts on learning and writing. Laurie Abkemeier, my agent, was instrumental in taking my rough ideas in a proposal and pushing me to develop something worthy of print. I thank Stephanie Hitchcock for editing the book and providing me with excellent feedback and suggestions. I'm also thankful to my friends and family who read over early drafts of the proposal and manuscript, helping the idea take shape. In particular, I would like to thank Zorica Tomovska, Vatsal Jaiswal, Tristan de Montebello, James Clear, Josh Kaufmann, Kalid Azad, and Barbara Oakley for their early feedback.

I would like to thank the wonderful people I met and interviewed while preparing for the book. I am grateful to Roger Craig, Eric Barone, Vishal Maini, Diana Jaunzeikare, Colby Durant, and Vatsal Jaiswal, who were kind enough to take time to help me fill in the details of their incredible stories. I want to thank many of the researchers who walked me through their findings and helped me understand the science of learning better. In particular, I want to thank K. Anders Ericsson for his patience as he helped me clarify many important points. In addition, I thank Robert Pool, Jeffrey Karpicke, Angelo DeNisi, Avraham Kluger, Jacqueline Thomas, and Michael Herzog for helping me understand the nuances of the science discussed in this book. I want to thank all the people who participated in my experiments with coaching ultralearning: Tristan de Montebello, Jeff Russell, Diana Fehsenfeld, Kate Schutt, Lissa Sherron, Joshua Sandeman, Keerthi Vemulapalli, Brittany Hsu, Shankar Satish, Ashima Panjwani, Ashfaq Alsam, Deepti Kannapan, and Ankita J.

Finally, I want to thank my parents, Douglas and Marian Young, both teachers, who taught me that learning is its own reward.

APPENDIX

Further Notes on My Ultralearning Projects

THE MIT CHALLENGE

Goal: Learn the material taught in MIT's undergraduate curriculum for computer science, using their freely provided materials and used textbooks

Method: Aim to pass all of the final exams (score over 50 percent, unless other information was provided) and complete the programming projects

Timeframe: October 2011 to September 2012

Notes and Discussion

It's important to note that what I ended up completing was not a facsimile of an MIT degree. Although I strove, whenever possible, to benchmark the overall curriculum covered and the intensity of evaluation, there were necessary departures from how an actual MIT student would have progressed through the same material.

At the level of the entire curriculum there were changes.

MIT's OpenCourseWare didn't offer options for humanities that I could grade myself at the time, so I swapped those for classes in economics. Lab-heavy classes for which I didn't have access to the equipment were substituted for pencil-and-paper theory classes I could do. MIT students were expected to carry out a thesis project. I didn't do that during my twelve-month study period, but for fun I did create a computer program that would allow someone to play Scrabble against a computer opponent shortly after my project had officially concluded. In evaluating the programming projects, I simply counted them as a success if they worked and performed the desired functions or were able to complete the accompanied testing suites.

For final exams, my default benchmark was to achieve at least 50 percent. I stuck to the official grading rubric whenever possible. When there were gaps (such as how to deduct points for arithmetic or algebraic errors on multistep problems), I used my judgment. The latter stage introduced some potential bias, so I decided to go back several years after completing the challenge and reevaluate all my exams using the strictest possible grading scheme (any mistake on a multipart question would make the entire question worth zero points; any incorrect result applied to further questions would make those questions also worth zero points). The outcome was that six of the thirty-three classes I had recorded as a "pass" would have been counted as a "fail" under this stricter schema. I don't believe this evaluation is the correct one, so I stand by my original evaluation of having passed those exams, but it is worth pointing out to show how much impact my subjective decisions had. A few classes had no final exams, so in those cases evaluation defaulted to assignments or midterm exams. Completing assignments was not a requirement to complete a class, however I did end up doing many of them as part of the learning process.

For more information on the challenge, such as course lists, materials used and scans of my exams, you can visit the challenge homepage: www.scotthyoung.com/blog/mit-challenge/.

THE YEAR WITHOUT ENGLISH

Goal: Learn Spanish, Portuguese, Mandarin Chinese, and Korean

Method: Avoid speaking English for the entire year, while traveling to Spain, Brazil, China, and South Korea (roughly three months each). I did this project along with Vatsal Jaiswal (who is also mentioned in chapter six).

Timeframe: September 2013 to August 2014

Notes and Discussion

Quantifying the level of proficiency reached in each language is a tricky task. There's a dual risk, both of exaggeration—implying a perfect level of fluency when that is probably a process requiring decades of immersion—and of downplaying. One person I talked to after the trip asked if I could "give a taxi driver directions" even though this task only requires a few hours of practice, not months. So with those difficulties in mind, I'll try to estimate the level that we reached:

Spanish: Here I believe both my friend and I reached a roughly B2 level after three months, meeting Benny Lewis's standard for fluency (although certainly not everyone's standard). At that level, we had little difficulty socializing for hours on any topic in Spanish, although certainly our accent, grammar and more formal speaking abilities were not at the level of a native speaker.

Portuguese: We were weaker in Portuguese than in Spanish, although not substantially so. The two languages share a common base, so there was much less to learn than there had been with Spanish. We could make friends and socializing, but not quite as effortlessly.

Mandarin Chinese: This marked the first big divergence in our abilities. I had really wanted to learn Chinese and had spent some time on flash cards prior to our trip to familiarize myself. My friend was less interested and struggled more. In the end, I wrote and passed the HSK 4 exam (the fourth in a six-level series of exams measuring Chinese proficiency) and I would say my Mandarin was decent, although more limited on advanced topics, where the vocabulary is completely different from English. My friend reached a lower-intermediate level, being able to speak comfortably and use tones but with less vocabulary.

Korean: In this language we both reached a lower-intermediate level, able to have conversations and get by in daily life, but on a more restricted range of topics. Part of this was the difficulty of the Korean language, but a bigger part was simply that as it was the fourth new language in a row, we were getting burned out.

Although we aimed to do most of our learning after arriving in each country, we did do some prior preparation for each. This was mostly listening to Pimsleur audio tapes and doing some flash cards. In generally we spent around twenty-five to fifty hours per language, although I spent more on Chinese (approximately one hundred hours) prior to arriving.

Those interested can see more about our project (including videos we put together showing our progress in each country), what we used to learn as well as unscripted interview to show roughly the level we reached in each language on the project homepage: www.scotthyoung.com/blog/the-year-without-english/.

PORTRAIT DRAWING CHALLENGE

Goal: To improve my ability to draw faces realistically
Method: Rapid feedback, techniques from various books and courses
Timeframe: July 2016

Notes and Discussion
This was a shorter project, taking one month and totaling one hundred hours of practice. In addition to the strategy of drawing quick sketches and comparing them by overlaying them on semi-transparent reference photos, I also greatly benefitted from the book *Drawing on the Right Side of the Brain*, and from Vitruvian Studio's Portrait Drawing class.

I've uploaded every drawing, sketch, and self-portrait I did, along with a more detailed discussion of what I used to learn on the project homepage: www.scotthyoung.com/blog/myprojects /portrait-challenge/.

FURTHER CHALLENGES

At the time of writing this book, the above three challenges are my main public ultralearning projects. However, I'm always learning new things, so as I do more public challenges, I'll post them here: www.scotthyoung.com/blog/my-projects/.

NOTES

Chapter I: Can You Get an MIT Education Without Going to MIT?

8   The Goethe-Institut, which administers: "Further Information," Goethe-Institut, https://www.goethe.de/en/spr/kup/prf/prf/gc2/inf.html.

9   "My first thought wasn't 'Wow'": Thanh Huynh, *Roger Craig—Knowledge Tracking,* filmed August 2011, YouTube video, 14:20, posted November 2011, https://www.youtube.com/watch?v=jmld3pcKYYA&t=1s.

10 "Everybody that wants to succeed at a game": "How One Man Played 'Moneyball' with 'Jeopardy!,'" National Public Radio, https://www.npr .org/2011/11/20/142569472/how-one-man-played-moneyball-with-jeopardy.

11 Spaced-repetition software is: Gary Wolf, "Want to Remember Everything You'll Ever Learn? Surrender to This Algorithm," *Wired,* April 20, 2008, https://www.wired.com/2008/04/ff-wozniak/?currentPage=all.

12 "You can simulate the game": Huynh, *Roger Craig—Knowledge Tracking.*

15 "incredibly endearing and beautiful": Patrick Hancock, "Review: Stardew Valley," Destructoid, March 7, 2016, https://www.destructoid.com/review -stardew-valley-345495.phtml.

16 "This is the type of person": "College Too Expensive? This Guy Just Finished a Four Year Computer Science Program in ONE Year Using Free MIT Material" (video), Reddit, https://www.reddit.com/r/videos/comments /10tk9j/college_too_expensive_this_guy_just_finished_a/.

21 Done without the benefit: Steve Pavlina, "Graduating College in 3 Semesters," December 4, 2005, https://www.stevepavlina.com/blog/2005/12 /graduating-college-in-3-semesters/.

21 Diana Jaunzeikare embarked on: Diana Jaunzeikare, "Personal PhD." https://diana.is/personal-phd.

22 "70–80+ hours each week": Tamu, "Independent Chinese Study: Review," Chinese-forums.com, https://www.chinese-forums.com/forums/topic/43939 -independent-chinese-study-review/.

22 Trent Fowler, starting in early 2016: Trent Fowler, *The STEMpunk Project* (Self-published, 2017).

Chapter II: Why Ultralearning Matters

27 "Average is over": Tyler Cowen, *Average Is Over: Powering America Beyond the Age of the Great Stagnation* (New York: Penguin, 2013).

28 The MIT economist David Autor: David H. Autor, Lawrence F. Katz, and Melissa S. Kearney, "The Polarization of the U.S. Labor Market," *American Economic Review* 96, no. 2 (May 2006): 189–94.

29 Tuition has increased far faster: Danielle Douglas-Gabriel, "College Costs Rising Faster than Financial Aid, Report Says," *Washington Post*, October 26, 2016,https://www.washingtonpost.com/news/grade-point/wp/2016 /10/26/college-costs-rising-faster-than-financial-aid-report-says/?utm_term =.72c95b4c86cb.

35 "a leading English-language novelist": Gareth Cook, "The Singular Mind of Terry Tao," *New York Times*, July 24, 2015, https://www.nytimes.com /2015/07/26/magazine/the-singular-mind-of-terry-tao.html.

Chapter IV: Principle 1—Metalearning: First Draw a Map

51 "Kuti paoka djalou": Linguistic Society of America, "'Monolingual Fieldwork Demonstration'—Daniel Everett," filmed July 2013, YouTube video, 1:16:27, posted September 2013, https://www.youtube.com/watch?v=sYp Wp7g7XWU.

52 What makes this feat particularly impressive: To avoid ruining the demonstration by using English, a language the other speaker might have been familiar with, Everett chose to phrase all his initial queries in the Pirahã language, spoken only by a remote people in the Amazon jungle of Brazil.

54 Over the last thirty years: The unusualness of this language has led to somewhat of a controversy in linguistics, with Dan Everett's claims about Pirahã's grammar at center stage in an attack on linguistic orthodoxy.

55 To see why metalearning is so: Jacqueline Thomas, "The Role Played by Metalinguistic Awareness in Second and Third Language Learning," *Journal of Multilingual and Multicultural Development* 9, no. 3 (1988): 235–46, https://www.tandfonline.com/doi/abs/10.1080/01434632.1988.9994334.

59 Determine if learning: Don't take this to mean that I think grad school is useless. The important thing to decide is whether it will really matter to you, depending on the job you want, the subject of your study, and the institution. My point isn't that grad school is a waste of time but rather that when making a decision involving so much time and cost, you'd better do the research first!

66 For example, one common recommendation: Victor Mair, "How to Learn Chinese and Japanese," Language Log, February 17, 2014, http://language log.ldc.upenn.edu/nll/?p=10554.

66 The literature on self-directed learning: George E. Spear and Donald W. Mocker, "The Organizing Circumstance: Environmental Determinants in Self-Directed Learning," *Adult Education Quarterly* 35, no. 1 (March 1, 1984): 1–10, https://journals.sagepub.com/doi/abs/10.1177/0001848184035 001001?journalCode=aeqb.

67 That led me to do: "Portrait Drawing—The Complete Online Course," Vitruvian Studio, https://vitruvianstudio.com/course/portrait-drawing/.

Chapter V: Principle 2—Focus: Sharpen Your Knife

70 Somerville explained, "she would have been contented": Mary Somerville, *Personal Recollections, from Early Life to Old Age, of Mary Somerville: With Selections from Her Correspondence* (London: Roberts Brothers, 1874), 23.

78 Hence, skilled performers: K. Anders Ericsson, *The Road to Excellence: The Acquisition of Expert Performance in the Arts and Sciences, Sports, and Games* (New York: Psychology Press, 2014), 25.

79 Similarly, the phenomenon: John Dunlosky, Katherine A. Rawson, Elizabeth J. Marsh, et al., "Improving Students' Learning with Effective Learning Techniques," *Psychological Science in the Public Interest* 14, no. 1 (January 8, 2013): 4–58, https://journals.sagepub.com/doi/abs/10.11771529100612453266.

83 "learn to let it arise": Susan L. Smalley and Diana Winston, *Fully Present: The Science, Art, and Practice of Mindfulness* (Philadelphia: Da Capo Lifelong Books, 2010), 59.

84 High arousal creates: A. E. Bursill, "The Restriction of Peripheral Vision During Exposure to Hot and Humid Conditions," *Quarterly Journal of Experimental Psychology* 10, no. 3 (August 1, 1958): 113–29.

84 Too much arousal, however: This inverse-U shape of arousal versus performance is known in psychology as the Yerkes-Dodson law.

84 More complex tasks: Daniel Kahneman, *Attention and Effort* (Englewood Cliffs, NJ: Prentice-Hall), 1973.

84 When doing a particularly creative task: Kalina Christoff, Zachary C. Irving, Kieran C. R. Fox, et al., "Mind-Wandering as Spontaneous Thought: A Dynamic Framework," *Nature Reviews Neuroscience* 17, no. 11 (2016): 718–31, https://www.nature.com/articles/nrn.2016.113.

85 In one experiment, sleep-deprived: Robert T. Wilkinson, "Interaction of Noise with Knowledge of Results and Sleep Deprivation," *Journal of Experimental Psychology* 66, no. 4 (November 1963): 332–37, https://psycnet.apa.org/record/1964–03490–001.

Chapter VI: Principle 3—Directness: Go Straight Ahead

88 Jaiswal leaves the offices: This is, in fact, the same Vatsal Jaiswal who joined me on my yearlong language learning project in chapter 1. These events took place a few years prior to that.

94 "Despite the importance": Robert Haskell, *Transfer of Learning* (Cambridge, MA: Academic Press, 2000), xiii.

95 In another study, college graduates were asked: James F. Voss, Jeffrey Blais, Mary L. Means, Terry R. Greene, and Ellen Ahwesh, "Informal Reasoning and Subject Matter Knowledge in the Solving of Economics Problems by Naive and Novice Individuals," *Cognition and Instruction* 3, no. 3 (1986): 269–302.

95 "in almost all the empirical work to date": Michelene T. H. Chi and Miriam Bassok, "Learning from Examples via Self-explanations," *Knowing, Learning, and Instruction: Essays in Honor of Robert Glaser* (1989): 251–82.

95 "students who receive honors grades": Howard Gardner, *The Unschooled Mind: How Children Think and How Schools Should Teach*, Basic Books (AZ), 2011.

95 "Researchers who rigorously evaluate training": John H. Zenger, "Great Ideas Revisited. The Painful Turnabout in Training. A Retrospective," *Training and Development* 50, no. 1 (1996): 48–51.

96 "Transfer is paradoxical": Wilbert J. McKeachie, "Cognitive Skills and Their Transfer: Discussion," *International Journal of Educational Research* 11, no. 6 (1987): 707–12.
103 Better graphics and sounds: Robert W. Proctor, and Addie Dutta, *Skill Acquisition and Human Performance* (Thousand Oaks, CA: Sage Publications, 1995).

Chapter VII: Principle 4—Drill: Attack Your Weakest Point

106 However, it was in the latter half: Benjamin Franklin, *The Autobiography of Benjamin Franklin* (New Haven, CT: Yale University Press, 2003).
106 world-changing consequences: Walter Isaacson, *Benjamin Franklin: An American Life* (New York: Simon and Schuster, 2003).
107 "written equally well": Ibid.

Chapter VIII: Principle 5—Retrieval: Test to Learn

119 What's more, he claimed: Robert Kanigel, *The Man Who Knew Infinity: A Life of the Genius Ramanujan* (New York: Simon and Schuster, 2016).
122 This is essentially the question: Jeffrey D. Karpicke, and Janell R. Blunt, "Retrieval Practice Produces More Learning than Elaborative Studying with Concept Mapping," *Science* 331, no. 6018 (February 11, 2011): 772–75, http://science.sciencemag.org/content/331/6818/772.
124 Minutes after studying something: Henry L. Roediger III and Jeffrey D. Karpicke, "The Power of Testing Memory: Basic Research and Implications for Educational Practice," *Perspectives on Psychological Science* 1, no. 3 (September 1, 2006): 181–210, https://journals.sagepub.com/doi/abs/10.1111/j.1745–6916.2006.00012.x?journalCode=ppsa.
124 Inevitably, students who were performing: Jeffrey D. Karpicke, "Metacognitive Control and Strategy Selection: Deciding to Practice Retrieval During Learning," *Journal of Experimental Psychology: General* 138, no. 4 (2009): 469–86, http://memory.psych.purdue.edu/downloads/2009_Karpicke_JEPGeneral.pdf.
125 One answer comes: Robert A. Bjork, "Memory and Metamemory Considerations in the Training of Human Beings," in *Metacognition: Knowing About Knowing*, ed. J. Metcalfe and A. Shimamura (Cambridge, MA: MIT Press, 1994): 185–205.
126 Delaying the first test: Jeffrey D. Karpicke and Henry L. Roediger III, "Expanding Retrieval Practice Promotes Short-Term Retention, but Equally Spaced Retrieval Enhances Long-Term Retention," *Journal of Experimental Psychology: Learning, Memory, and Cognition* 33, no. 4 (July 2007): 704–19, http://memory.psych.purdue.edu/downloads/2007_Karpicke_Roediger_JEPLMC.pdf.
126 However, if you delay the test: Herbert F. Spitzer, "Studies in Retention," *Journal of Educational Psychology* 30, no. 9 (December 1939): 641–56, https://www.gwern.net/docs/spacedrepetition/1939-spitzer.pdf.
126 An interesting observation: Chunliang Yang, "Enhancing Learning and Re-

trieval: The Forward Testing Effect," PhD diss., University College London, 2018.

Chapter IX: Principle 6—Feedback: Don't Dodge the Punches

136 "It's not going to be": Kelefa Sanneh, "Chris Rock, the Duke of Doubt," *New Yorker*, November 10, 2014, https://www.newyorker.com/magazine /2014/11/10/duke-doubt.

137 Many medical practitioners get worse: Anders Ericsson and Robert Pool, *Peak: Secrets from the New Science of Expertise*, (New York: Houghton Mifflin Harcourt, 2016).

138 In a large meta-analysis, Avraham Kluger: Avraham N. Kluger, and Angelo DeNisi, "The Effects of Feedback Interventions on Performance: A Historical Review, a Meta-analysis, and a Preliminary Feedback Intervention Theory," *Psychological Bulletin* 119, no. 2 (1996): 254–84, https://psycnet.apa .org/record/1996–02773–003.

142 In one study, feedback: Michael H. Herzog and Manfred Fahle, "The Role of Feedback in Learning a Vernier Discrimination Task," *Vision Research* 37, no. 15 (August 1997): 2133–41, https://ac.els-cdn.com/S0042698997000436/1-s 2.0-S0042698997000436-main.pdf?_tid=9e63a472–9df4–43fa-a165–7ff 3daa4ddd2&acdnat=1551035784_e6ebf10b08703a5479c3abbf649b5320.

144 "The best feedback is *informative*": Maria Araceli Ruiz-Primo and Susan M. Brookhart, *Using Feedback to Improve Learning* (New York: Routledge, 2017), 128.

146 James A. Kulik and Chen-Lin C. Kulik review the literature: James A. Kulik and Chen-Lin C. Kulik, "Timing of Feedback and Verbal Learning," *Review of Educational Research* 58, no. 1 (1988): 79–97.

147 Expertise researcher: K. Anders Ericsson, Ralf T. Krampe, and Clemens Tesch-Römer, "The Role of Deliberate Practice in the Acquisition of Expert Performance," *Psychological Review* 100, no. 3 (1993): 363–406, https:// psycnet.apa.org/record/1993–40718–001.

147 In those studies, however: Wendy Jaehnig and Matthew L. Miller, "Feedback Types in Programmed Instruction: A Systematic Review," *Psychological Record* 57, no. 2 (2007): 219–32.

Chapter X: Principle 7—Retention: Don't Fill a Leaky Bucket

153 French, with its gendered nouns: Corazon Miller, "How Kiwi Nigel Richards Won French Scrabble Championship," *New Zealand Herald*, July 22, 2015, https://www.nzherald.co.nz/lifestyle/news/article.cfm?c_id=6&object id=11485116.

154 "Nigel, since you're no good at words": Zeba Sultan, "Nigel Richards— An Enigma," The paladin speaks . . . http://vivaciouspaladin.blogspot.com /2013/05/nigel-richardsan-enigma.html.

155 "When I see you, I can never tell": Stefan Fatsis, "Nigel Richards Article," Scrabble Study Log, http://scrabblestudylog.blogspot.com/2009/08/nigel -richards-article-by-stefan-fatsis.html.

156 He politely declined: Tim Hume, "A Way with Words," *Sunday Star-Times*,

June 6, 2010, http://www.stuff.co.nz/sunday-star-times/features/3778594
/A-way-with-words.

157 "The cycling helps": Fatsis, "Nigel Richards Article."

158 "It's hard work": Daniel Stembridge, "Meeting Nigel Richards," Mindsports
Academy, https://www.mindsportsacademy.com/Content/Details/2133?title
=meeting-nigel-richards.

158 "I'm not sure there is a secret": OgilvyBroadcast, "World Scrabble Championships 2011," filmed October 2011, YouTube video, 1:51, posted October
2011, https://www.youtube.com/watch?v=EZE_olsi-pM&t=1m46s.

159 "Physicians with more experience": Niteesh K. Choudhry, Robert H.
Fletcher, and Stephen B. Soumerai, "Systematic Review: The Relationship
Between Clinical Experience and Quality of Health Care," *Annals of Inter-
nal Medicine* 142, no. 4 (2005): 260–73, https://annals.org/aim/fullarticle
/718215/systematic-review-relationship-between-clinical-experience
-quality-health-care.

163 This seems especially likely: Joyce W. Lacy and Craig E. L. Stark, "The
Neuroscience of Memory: Implications for the Courtroom." *Nature Reviews
Neuroscience* 14, no. 9 (September 2013): 649–58, https://www.ncbi.nlm.nih
.gov/pmc/articles/PMC4183265/.

166 The authors of a popular study guide: Peter Wei and Alex Chamessian,
Learning Medicine: An Evidence-Based Guide (Self-published, 2015).

167 procedural skills, such as: Jong W. Kim, Frank E. Ritter, and Richard J.
Koubek, "An Integrated Theory for Improved Skill Acquisition and Retention in the Three Stages of Learning," *Theoretical Issues in Ergonomics Science*
14, no. 1 (2013): 22–37.

170 Overlearning is a well-studied: James E. Driskell, Ruth P. Willis, and Car-
olyn Copper, "Effect of Overlearning on Retention," *Journal of Applied
Psychology* 77, no. 5 (1992): 615–22, https://psycnet.apa.org/record/1993
-04376-001.

172 One study of algebra students: Harry P. Bahrick and Lynda K. Hall, "Lifetime Maintenance of High School Mathematics Content," *Journal of Experimental Psychology: General* 120, no. 1 (1991): 20–33, http://citeseerx.ist.psu
.edu/viewdoc/download?doi=10.1.1.1020.7785&rep=rep1&type=pdf.

173 Rajveer Meena, the Guinness World Record: "Most Pi Places Memorised,"
Guiness World Records, http://www.guinnessworldrecords.com/world
-records/most-pi-places-memorised.

Chapter XI: Principle 8—Intuition: Dig Deep Before Building Up

177 "a magician of the highest caliber": James Gleick, *Genius: The Life and Science
of Richard Feynman* (New York: Vintage, 1993), 10.

178 "He's the only guy": Richard P. Feynman and Ralph Leighton, *"Surely You're
Joking, Mr. Feynman!": Adventures of a Curious Character* (New York: Random
House, 1992), 133.

180 "I happened to know": Ibid., p. 193.

181 "I had a scheme": Ibid., p. 85.

182 In a famous study, advanced PhDs: Michelene T. H. Chi, Paul J. Feltovich, and Robert Glaser, "Categorization and Representation of Physics Problems by Experts and Novices," *Cognitive Science* 5, no. 2 (April 1981): 121–52, https://onlinelibrary.wiley.com/doi/pdf/10.1207/s15516709cog0502_2.

182 Another study, this time: William G. Chase and Herbert A. Simon, "Perception in Chess," *Cognitive Psychology* 4, no. 1 (January 1973): 55–81, http://citeseerx.ist.psu.edu/viewdoc/download?doi=10.1.1.601.2724&rep=rep1&type=pdf.

183 Researchers have estimated: Fernand Gobet and Herbert A. Simon, "Expert Chess Memory: Revisiting the Chunking Hypothesis," *Memory* 6, no. 3 (1998): 225–55, https://pdfs.semanticscholar.org/d11f/079a1d6d3147abbb78 68955a6231f4a5ba5b.pdf.

185 "If [he] had said": Feynman and Leighton, *"Surely You're Joking, Mr. Feynman!,"* 21.

186 Feynman told a story: The work, which won the pair the Nobel Prize, demonstrated that the universe we live in is not mirror-image symmetrical. That is to say, there are certain physical processes that look different in a mirror version. At the time, it was an enormous surprise to physicists, who had assumed that this symmetry existed. Ibid., 249.

187 One of Einstein's earliest: Walter Isaacson, *Einstein: His Life and Universe* (New York: Simon and Schuster, 2008).

187 "illusion of explanatory depth": Rebecca Lawson, "The Science of Cycology: Failures to Understand How Everyday Objects Work," *Memory & Cognition* 34, no. 8 (2006): 1667–75, http://gearinches.com/misc/science-of -cycology.PDF.

188 Feynman's and Einstein's approach: The artist and designer Gianluca Gimini plays on this concept by designing bicycles that look as people think they ought to (but that of course don't work). You can see some of his creations at gianlucagimini.it/prototypes/velocipedia.html.

189 In one study of this effect: Fergus I. M. Craik and Robert S. Lockhart, "Levels of Processing: A Framework for Memory Research," *Journal of Verbal Learning and Verbal Behavior* 11, no. 6 (December 1972): 671–84, http://wixtedlab.ucsd.edu/publications/Psych%20218/Craik_Lockhart_1972.pdf.

189 Those who processed the words: Thomas S. Hyde and James J. Jenkins, "Differential Effects of Incidental Tasks on the Organization of Recall of a List of Highly Associated Words," *Journal of Experimental Psychology* 82, no. 3 (1969): 472–81, https://people.southwestern.edu/~giuliant/LOP_PDF/Hyde 1969.pdf.

190 The Dunning-Kruger effect occurs: Justin Kruger and David Dunning, "Unskilled and Unaware of It: How Difficulties in Recognizing One's Own Incompetence Lead to Inflated Self-Assessments," *Journal of Personality and Social Psychology* 77, no. 6 (December 1999): 1121–34, https://pdfs.semantic scholar.org/e320/9ca64cbed9a441e55568797cbd3683cf7f8c.pdf.

191 "Some people think": Feynman and Leighton, *"Surely You're Joking, Mr. Feynman!,"* 244.

193 I had this uneasy feeling: Ibid., 281.

194 With my textbook at my side: You can view my notes here: https://www
.scotthyoung.com/mit/photogrammetry.pdf.

194 To get a better handle: You can view my notes here: https://www.scotth
young.com/mit/grid-accel.pdf.

196 "I got it down": Ibid., 141.

Chapter XII: Principle 9—Experimentation:
Explore Outside Your Comfort Zone

198 "You started too late": Steven W. Naifeh and Gregory White Smith, *Van Gogh: The Life* (New York: Random House, 2011), 260.

198 "We considered his work": Ibid., 514.

199 sold for more than $82 million: Judd Tully, "$82.5 Million for van Gogh; Japanese Buyer Sets Art Auction Record," http://juddtully.net/auctions/82–5
-million-for-van-gogh-japanese-buyer-sets-art-auction-record/.

201 "devoured these big books": Naifeh and Smith, *Van Gogh*, 214.

202 "Scarcely any color is not gray": Ibid., 333.

207 growth mindset: Carol S. Dweck, *Mindset: The New Psychology of Success* (New York: Random House, 2008).

209 It may also dispel: I had my own experience when trying to write this book. As part of my process, I reread many other books whose style I wanted to emulate. In doing this, a thing that surprised me was that many such books had far fewer citations than I remembered, the "seriousness" of a book being mostly a matter of tone, not of scholarship.

211 Scott Adams, the creator of *Dilbert*: Scott Adams, "Career Advice," Dilbert
.Blog, July 20, 2007, http://dilbertblog.typepad.com/the_dilbert_blog
/2007/07/career-advice.html.

Chapter XIV: An Unconventional Education

234 "Grandmasters don't like to lose": The source I've been able to track down seems to be here: Shelby Lyman (02–08–1987), "Younger Sisters Are Also Proficient," *Sunday Telegraph* 1 (45).

234 "three or four great chess prodigies in history": F. Lidz, "Kid with a Killer Game," *Sports Illustrated* 72, no. 6 (1990): 8–8.

234 "She has fantastic chess talent": Ibid.

235 "I was playing the World Champion": *Chess Life* 50, (no. 7–12): 647.

235 "How could you do this to me?": Leonard Barden, "Sweet Revenge for Kasparov's Opponent," *Guardian*, September 11, 2002, https://www.the
guardian.com/world/2002/sep/11/3.

235 "I think a girl of her age": Dirk Jan ten Geuzendam, "Finding Bobby Fischer: Chess Interviews by Dirk Jan ten Geuzendam," *Alkmaar, the Netherlands: New in Chess* (1994), 203.

236 "A genius is not born": Peter Maass, "Home-Grown Grandmasters," *Washington Post*, March 1992.

236 "[W]hen I looked at the stories": Linnet Myers, "Trained to Be a Genius,

Girl, 16, Wallops Chess Champ Spassky for $110,000," *Chicago Tribune*, February 1993.

238 "Women are able": Patricia Koza, "Sisters Test Male Domination of Chess," *Mohave Daily Miner,* November 1986.

239 "The Polgárs showed": G. K. Kasparov and Mig Greengard, *How Life Imitates Chess: Making the Right Moves, from the Board to the Boardroom* (New York: Bloomsbury, 2008).

240 "Have they been educated": László Polgár, *Raise a Genius!* (Vancouver: self-published, 2007), 97, https://docplayer.net/64270951-Raise-a-genius -by-laszlo-polgar-original-edition-laszlo-polgar-nevelj-zsenit-budapest -interviewer-endre-farkas.html.

243 "starting from 4–5": Ibid., 33.

243 "play is not the opposite": Ibid., 20.

244 "Following a number": Ibid., 16.

244 "We should make": Ibid., 51.

244 "One thing is certain": Ibid.

244 "one of the most important": Ibid., 36.

248 "opening preparation was not": Judit Polgár, *How I Beat Fischer's Record* (Glasgow: Quality Chess UK Ltd, 2012), 11.

INDEX

ABOUT THE AUTHOR

SCOTT H. YOUNG is a writer who undertakes interesting self-education projects, such as attempting to learn MIT's four-year computer science curriculum in twelve months and learning four languages in one year. He lives in Vancouver, Canada.